RESOURCE CONSERVATION AND FOOD SECURITY

RESOURCE CONSERVATION AND FOOD SECURITY

An Indian Experience

Volume II

Edited by
TAPESHWAR SINGH

CONCEPT PUBLISHING COMPANY, NEW DELHI-110059

ISBN : 81-8069-100-4 (Set)
81-8069-102-0 (Vol. II)

First Published 2004

Published and Printed by
Ashok Kumar Mittal
Concept Publishing Company
A/15-16, Commercial Block, Mohan Garden
NEW DELHI-110059 (India)

Phones : 25351460, 25351794
Fax : 091-11-25357103
Email : publishing@conceptpub.com

CONTENTS

SECTION-II
SUSTAINABLE AGRICULTURAL DEVELOPMENT AND FOOD SECURITY

VOLUME-II

SECTION-V
ENVIRONMENTAL CHANGE AND IMPACT

SECTION-VI
SOCIO-CULTURAL ASPECTS

INTRODUCTION

Tapeshwar Singh

In India nearly 70 per cent of population directly or indirectly depends on land for their livelihood. Sustained and equitable growth in agriculture is, therefore, crucial to the food security and sustainable development. Economic growth and evironmental preservation are no any more opposing objectives. Environment friendly economic growth has become a necessity. The day of frontier economy when abundant natural resources were available to propel economic growth and raise standard of living are over. We have reached a stage when the concern is with the sustainable development meaning thereby not only the development of the present but also of the future. Thus the focus has moved from sastained growth' to 'sustainable development'. Sustainable implies a notion of equilibrium, an equilibrium that simultaneously satisfy the needs of development as well as resource conservation. Technology has helped us to harness the forces of nature thereby enhancing our capacity to improve living standard and quality of life. However, our past achievements in raising resources efficiency and productivity should not delude us into a thinking that any constraint can be overcome. It is known that technology has proved to be double edged sword in many cases. Green revolution technology for example, has been very successful in achieving spectacular results in food grain production during the last three decades. However, signs of fatigue in the natural resources have already emerged and have unleashed various agro-ecological problems. It has badly damaged the natural resource base of the country. Great apprehensions are, therefore, voiced about the stability of enhanced productivity and environmental safety as also the grave limitation to feed the teeming millions in 21st century. Realizing the dangerous implications of the emerging scenario particularly rapid depletion and degradation of natural resources and considering the paramount importance of the inevitable prudent use of natural resources an International Symposium on Land Degradation : New Trends Towards Sustainable Agriculture simultaneously with the Commonwealth Geographical Bureau Food Security Workshop was organized during April 7-12, 2002 at Ghaziabad to take stock of the work done in the past and to formulate future strategies for efficient integrated management of the natural resource for sustainable agriculture. The workshop and symposium were organized by the Department of Geography, MMH College (Ch. Charan Singh University), Ghaziabad. It was cosponsored by State Land Use Board, Planning Department, Government of Uttar Pradesh; Council for Scientific and Industrial Research; Indian Council of Social Science Reasearch; National Informatics Centre (Minisry of Communication and Information Technology, Govt. of India); Department of Science and Technology, Government of India; Bhoovigyan Vikas Foundation and Commonwealth Geographical Bureau. It was inaugurated by Dr. Panjab Singh, the then Secretary, Department of Agricultural Research and Education and Director General of Indian Council of Agricultural Research with Shri Som Pal, Member, Planning Commission in the Chair. Around 225 scientists and other interested in judicious use of natural resources representing different universities and organizations from India and abroad participated. In total 65 oral and 15 poster presentations were made. Some of the participants could not attend but sent their contributions.

Land degradation under increasing abiotic and biotic stresses has been increasing sharply in the last two-three decades and has become central issue to our developmental strategies. The five-day long proceedings and deliberations were particularly concerned with the contentious issues like (i) monitoring land quality and sustainable agriculture; (ii) Land use ethics and social and evironmental implications of land degradation—social relevance of soils function and natural resource endowment and biodiversity, climate change etc.; (iii) Land reclamation strategies and tchniques—land use, planning and prevention of degradation, sustainable land management : soil and water conservation and management; (iv) Public policy to achieve sustainable land use; (v) land degradation assessment—biotic and abiotic indices of land degradation, early warning indicators, tools (GIS, DEM, Simulation Models, and Decision Support Systems), rehabilitation of degraded land, use of traditional knowledge, early warning system and database, GIS application and techniques (application of tools and methods to assess the impact of various land use scenarios) and increasing vulnerability of land degradation risks, simulation and inerdisciplinary cooperaition and assistance to get from different components of planning and implementation and (vi) *The* land resource base : vulnerability; degradation risks to land submitted to different land usė pressure, resilience-potential of misused/degraded land to naturally recovered to its orginal condition. The symposium also focused on the impacts of natural disaster such as drought and floods and the impact of environmental changes on land use and cropping pattern in food security context. Learned speakers expressed serious concerns on surfacing noticeable stagnancy in food output in some cropping systems. In the contemporary state of science and development it is very important to improve mutual understanding of natural and social scientists, policy-makers and implementing agencies as well as better feeling of scientific understanding of land resources for addressing these issues.

There is no space to present excerpts of all the papers presented, therefore, I would like to concentrate only on some introductory presentations given by Dr. Panjab Singh, Seceretary, Dapartment of Agricultural Research and Education and Director General of Indian Council of Agricultural Research, Shri Som Pal, Member, Planning Commission; Dr. Victor Roger Savage, President of Commonwealth Geographical Bureau and Head Department of Geography, National University of Singapore; Dr. Tony Binns and Dr. Don Funnell of Sessex University, London; Dr. D.R. Sikka, Former Director, Indian Institute of Tropical Meteorology, Pune; Dr. Jagdish Singh, former Head, Department of Geography, Gorakhpur University; Dr. S.N. Das, Chief Soil Scientist, All Indian Soil and Land Use Survey; Dr. B.B. Singh, Ex-Head, Department of Geography, M.M.H. College; Dr. M. Moni, Deputy Director General, National Informatics Centre, Ministry of Communication and Information Technology, Government of India; and Dr. Tapeshwar Singh, Convener, CGB Worshop and Symposium.

Shri Som Pal, Member, Planning Commission stressed the need for solving the food security problems at both macro as well as micro-levels. At national level we have become food self-sufficient but at households level, yet not successful in solving the problems of poverty, hunger and malnutrition. It is a big challenge before the country and has to be solved for attaining universal food security. Thus, food security should be assured for those vulnerable groups who are food-insecure and resource-poor, and are unable to transact in the market. While talking about sustainable agricultural development Shri Som Pal lamented that land degradation is very serious problem and said that about 57 per cent area in the country is potentially exposed to various degradation forces such as water and wind erosion, floods, acidification, waterlogging, sodicity, shifting cultivation, ravines and gullies, riverine and torrent land etc. The Indian agriculture is running with net negative balance of plant nutrient at the rate of 8-10 million tonnes a year. The health of Indian soil is becoming poorer day by day and second generation problems of green revolution such as imbalance in micronutrient are surfacing. Depletion of groundwater and side by side waterlogging, alarming loss of biodiversities, and desertification are threatening the very sustainability of the agroecosystem. He called upon geographers to come forward with feasible solutions to overcome these challenges.

Tony Binns speaking on 'food security: changing the focus' presented various definitions of food security and elaborated extensively the different approaches in historical perspective—modern and post-modern approaches in development, post-modernism and food security, case study: issues concerning food security in Southern Africa, understanding indigenous livelihood systems: ensuring food security, future food security—an action plan etc. He particularly focused on the need of ACCESS (Food entitlement) and CONTROL over resources. Dr. Binns discussed in detail various facets of the coping strategies of food and livelihood on various scales—global, national, community, household and individual. The discussion on a 'food first' perspective to a 'livelihood security' perspective since 1985 drew good attention and appreciation. Lamenting on the action plan Dr. Binns suggested various strategies for attaining sustainable food security such as focus on supplying vulnerable people and households with secure access to food. Individual and household needs should take precedence over national food self-sufficiency and self-reliance. need for poverty-reducing economic growth to generate secured livelihoods, (with buffers against destitution) and buoyant national economies to provide jobs and, where necessary purchase of food, in agricultural sector—need for generating labour-intensive strategies which emphasise job creation and better incomes for poorest groups, balance between food and cash crops, emphasis on long-term comparative advantage rather than self-sufficiency for its own sake, efficient food marketing to store and distribute food at reasonable prices to all parts of the country in all seasons and in all years, improved famine preparedness and response, especially at local level, locally-based stock for relief distribution, programmes for vulnerable group feeding and rehabilitation etc. Dr. Binns also argued to recognize causes of food-insecurity, situation and strategies, providing households and individuals with choices, which contribute to self-determination and autonomy in developing livelihood strategies. Food security planning should follow a 'process' rather than a blueprint approach of decentralization, bias to action over planning, encourages risk taking and innovation, multi-disciplinary and multi-sectoral planning teams and strong international support in crisis management.

Don Funnell explained the significance of Geographic Information System in scientific studies, while presenting a paper on 'Food Security and Geospatial Analysis'. He presented comprehensive analysis of GIS application for resource appraisal and solving the food problem at different spatial levels (national, regional and local). He stressed the need to study the link between food availability and nutritional well-being which is not automatic, the impact of new agenda such as globalisation, biotechnology etc., whether food security remains a useful policy concept and the role of food aid in providing food security. Dr. Funnell observed that there has been a rapid growth in the development of affordable technologies associated with the provision of data which are flagged off to assist in the decision making process. Easily available computing power, cheap and relatively accurate GPS equipment, access to the Internet by more and more people and agencies and the development of PPGIS (sometimes known as GIS2) have widened the scope of accuracy. Information and Communication Technology (ICT) have been developed which promised not just improved methods of achieving existing aims but, arguably more important, providing the capacity for isolated and disenfranchised communities to participate in the whole process of managing food security in a manner that offers new ways of thinking about the process. The food security debate is becoming more nuanced in the light of a better understanding of the food and mutritional system but that ICT, especially geospatial technologies have become increasingly significant in tackling these problems.

Geospatial analysis must be involved in the following processes :

(i) *Generating data* : In this category falls the wide range of information gathering technologies ranging from 'high tech' use of satellite imagery and the use of GPS, through structured data collection by formal agencies.

(ii) *Handling data* : The particular forms in which data are categorised and stored.

(iii) *Interpreting data* : Food security policy usually requires the monitoring of the flows of food through an agricultural year. Depending upon the focus of attention concerning vulnerabilities. In some cases food vulnerability is not a function of gross supply or even of price *per se* but of weaknesses in other aspects of infrastructure such as transport, especially in the face of natural disasters such as earthquakes or floods. Data interpretation, therefore, is the process of using geospatial information.

(iv) *Developing policy* : National, regional, and local scales of policy formulation only really make sense when it is possible to obtain and analyse data which itself is available at these scales, too often, shortage of spatially referenced data at each of these scales.

(v) *Monitoring* : ICT can play a significant role in this area. One of the problems with most conventional data gathering procedures concerns the infrequency of data availability. Traditional data collection procedures is also time consuming hence using rapid appraisal techniques combined with input from satellite monitoring may help to update information continuously meanwhile checking spot anomalies.

Delivering the Valedictory Address Dr. D.R. Sikka, a renowned meteorologist underlined the vulnerability of the Indian agriculture as well as farmers, in view of global warming and changing climate and ecology. He laid stress on evolving adaptation strategies and resource-conserving techniques for sustainable farm development and improved productivity in the light of the current climate models prediction of a global warming of about 1.4 to 5.8 degree centigrade over the present century which is likely to have an impact on Indian agriculture resulting in up to 30 per cent drop in crop production by 2050. This is likely to have adverse impact particularly on rice and wheat yields. Eventhough there are some uncertainties in model prediction, it is important to keep these projections in view while formulating policies. South Asia, with its large population, low incomes, fragile ecosystems and dependence on agriculture and allied activities is particularly vulnerable to climate change. Dr. Sikka observed that projection of reduction in crop yields in our country were of grave concern, taking into consideration the growing food demands on account of rapidly increasing population. He also underlined the need to understand the social, environmental and ethical implications of land degradation and also warned against the danger of corporatisation of land in India and stressed the need to safeguard the interest of the small farmers from 'big sharks'. He also observed 'if we have to sustain agriculture we must have a proper database and information on land quality and all other determinants of agriculture'.

Jagdish Singh stressed the need of adopting the organic farming system (slowly and cautiously) that holds promise for the 'right of nourishment' of the masses and sustainable food security as the fatigues of Green Revolution have already emerged and chemical input-based farming is petering out without fully covering all crops and regions of the country. Under such a backdrop we must evolve a new paradigm of sustainable agricultural development based on job-led economic growth rooted in the principles of ecology and equity.

S.N. Das considered watershed management as the most advance integrated approach for sustainable development and management of our natural resources. Developmental planning on watershed basis evolves eco-friendly land management system to meet the needs of ever-increasing human and animal population and mounting biotic and abiotic stresses on the environment and calls for realistic planning of soil and water conservation measures and land use management in the sense of economic feasibility and people's acceptability. Watershed management involves detailed inventory of database of key natural resources—soil and water and their interactive analysis with socio-economic condition and environment. Application of modern technologies like space technology, bio-technology and information technology suitably integrated with the environment to develop a cost-effective, energy-efficient and environmentally benign management system which can make development sustainable over time. Sustainability has many aspects—ecological, socio-economic and cultural and all have to be integrated.

B.B. Singh in his presentation on 'Food Security Dilemma' observed that food security is not the function of mere food production alone but it is more a question of economic access to food by all sections of society, particularly the most vulnerable one. Food security without livelihood security is insurmountable. He suggested vigorous employment generation in the farm and off-farm sectors to enable the resource-poor to have direct access to income as an integral strategy for ensuring food security, promotion of employment through rural industrialization and giving immediate priority for creating opportunities for value-addition integrating production with processing and marketing, and full utilization of territorial resources and production potential. As per government claims the rate of poverty in India has come down to 26 per cent over 5 decades of Independence and planned development. However, the absolute number of the poor remains almost equal to the size of total population of the country at the time of independence. Dr. Singh, therefore, emphasized the need of re-defining 'food security' and arguably suggested to establish a separate Ministry at the Centre and the States to deal exclusively with the persistent problem of food insecurity, unemployment and poverty.

M. Moni observed that at the end of 20th century, the agricultural enterprises stand at the technological crossroads. The existence (or absence) of favourable natural resources can facilitate (or retard) the process of economic development. India is endowed with regional diversities for its uneven economic and agricultural development. Out of total geographical area of 329 Mha of the country nearly 265 Mha represent varying degrees of potential for biological production. More than 57 per cent of land suffers from different types of degradation and it is a very serious environmental problem in India in a acceptable technology management framework in the prevailing socio-economic infrastructure. Land degradation inversely affects the land productivity, food security, climate and environmental security, and eventually the quality of life.

Emergence of Information Technology on the national agenda in India and announcement of IT policies by the respective State Governments have recognized the convergence of core technologies and E-Governance as the tool for sustainable development and globalization of economy. The fusion of Information and Communication Technology, Biotechnology, Environment Technology, GIS and Remote Sensing Technology are expected to bring in the desired S & T inputs for sustainable development. As a step towards reaching technology to the Small Holders (Resource-poor-farmers) and also to usher in agricultural reforms through ICT in the country, the paper discussed an *Informatics led development Strategy—from digital divide to digital opportunities* for sustainable agricultural and rural development, the need to develop a comprehensive Agricultural Resources Informatics using Geomatics Technology in districts with public funding, and also suggests need to develop metadata and application of Open GIS model for optimal utilization of agricultural resources in India.

While introducing the themes of the Workshop and Symposium Dr. Tapeshwar Singh observed that economic growth and environmental preservation are no more opposing objectives. Environment friendly economic growth has become a necessity. The days of frontier economy when abundant resources were available to propel economic growth and raise living standards are over. Action is needed to prevent degradation and regenerate degraded resources. We urgently need positive steps to extend the forest cover, rejuvenate degraded lands and other natural resources especially two key resources such as land and water. Ecological improvement can certainly condition sustainable development. The use of geomatics (GIS, Remote Sensing, GPS) is very significant in geographical studies, hence be incorporated at all levels from primary to higher education. A renaissance in geographical thinking is necessary to maintain the Domain of Geography as a science. We must develop a vision for achieving the ecologically sustainable agriculture through income enhancing, productivity growth promoting, resource conserving and environment friendly practices, policies and technologies that will conform to the emerging paradigms of efficiency, sustainability and environmental security for reducing poverty and achieving sustainable food security.

Recommendations

The three-day long discussions and exchange of ideas on up-to-date information by the delegates on the main theme and sub-themes of the Symposium across the spectrum of the Earth Sciences, Environment, Agriculture, Information Technology, Meteorology and Social Sciences concluded with the following recommendations :

- Declining soil productivity in India is threatening the food grain production. The challenge before the scientific community to combat poverty and hunger can be achieved through combating land degradation and at large soil and water conservation and management. Excessive use of land resources by rapidly growing population and diversion of prime agricultural land for other uses should be prevented and brought to a minimum.
- The land use policy should be optimizing returns on a long-term basis rather than meeting short-term requirement. The emphasis should be on the management of land resources on watershed basis. Integrated and holistic development of rainfed area will have to be promoted by conservation of land and water with involvement of watershed community.
- The potential ability of the system to meet firture food need must be measured against constraints being imposed by limited resources, mounting degradation and unfavourable policies since these factors will hamper the future growth and development.
- Farming will have to be knowledge-intensive, using information from Remote Sensing, Geographical Information System (GIS), Global Positioning System (GPS) and information and computer techniques.
- With necessary caution to prevent unforeseen problems associated with biotechnology the adoption of this technology will play an important role in strengthening food security in 21st century.
- The gene revolution, the ecotechnology revolution and the communication revoluton have to be included as the main components of resource management and the national agricultural policy for sustainable development of agriculture in future.
- Soil, water, biodiversity and climate-basic natural resources for agricultural growth and development, are subjected to

various forms of deteriorating influences and indiscriminate exploitation, hence needs conservation and management, particularly under the threat of anticipated climate change scenarios.

- In fact, the complex and intrinsic relationship between environment and agricultural development has received insufficient attention, therefore, improvement of environmental health for the goal of achieving sustainable agricultural development and food security, should be central to planning.
- A second green revolution is needed particularly in those areas, which have already been benefited but have also suffered environmental problem and stagnancy in food output under the prevailing biophysical conditions, and for environmentally fragile areas and people. But without repeating the disadvantages of the first revolution. However, green revolution should also cover all the areas left out from the benefit so far. It needs to aim both at increasing yield and at preserving and developing the environmental base. Therefore we must strive for promotion of sustainable green revolution and curbing of on-going geed revolution.
- Value addition to the agricultural products should be made one of the important components of sustainable agriculture.
- The components of climate change and variability will have to be incorporated in formulation of development policy and planning, particularly as water may become a scarce commodity. Since modern agriculture is water-intensive activity, there should be proper and judicious exploitation and management of water for sustainable agriculture.
- The Symposium strongly recommended to have a proper database on land quality and all other determinants of agriculture for sustainable agricultural development, particularly on subregional basis where the adverse impacts are felt.
- We must evolve a mechanism to ensure food security to every citizen. The persistence of hunger, despite enough food production can be combated only through economic access to food and recognition of entitlement right to food security

particularly to the most vulnerable sections. To achieve these objectives, a Hunger-Free Area Programme and National Food Security Act should be initiated for universal food security. It is also recommended that a separate 'Food Security Ministry' be established at Centre and State level to achieve the objectives.

- Finally, our long-term agricultural policy should be ecologically sound, economically viable, socially relevant and sustainable for different agroclimatic systems keeping in view the needs of the present as well as the future generation.

In view of the concerns expressed by the experts on the significance of the issue and need of update information, this volume is undertaken to be published. This volume contains a series of commissioned articles by experts in various fields, ranging from natural scientists to socio-economists and remote sensing and geographic information system, in consonance with the themes of the workshop and symposium. This comprehensive scientific information on wide ranging issues—soil and water resources and their depletion and degradation, weather and climate, agro-bio-diversity, cropping system, integrated resource management, sustainable agriculture and food security, demography, remote sensing and GIS, organic farming and agribusiness, science and technology for food security, land use and land cover, drought and flood management, bio-technology, impact of environmental changes on plants, global environment, sami-arid and arid ecosystem, hill ecology etc.

The 51 contributions published in this document pertain to leading policy makers, planners, technologists and scientists, economists and geographers. They have been brought together in this book to help define shared perceptions on the evironmental issue and to further sustain and re-examine resource conservation polices needed for a long-term agenda for action in the era of knowledge. It is necessary for a more determined and drastic measures towards achieving sustainabitlity. The aim of the publication is not only to discuss and help set out the facts clearly, but also to encourage public awareness and to achieve a desired political effect.

The paper included in this volume are divided into six major sections :

Section I :	Land and Water Resource Management
Section II :	Sustainable Agricultural Development and Food Security
Section III :	Weather and Climate
Section IV :	Geo-spatial Technology Application
Section V :	Environmental change and Impact
Section VI :	Socio-cultural Aspects

Nineteen papers included in Section I discuss the 'Land and Water Resource Management'. Majority of the papers have expressed great concern on the ecosystem sustainability. Over-exploitation beyond the resource capability or an imbalanced use or/and misuse not in harmony with the local condition in an attempt to augment rapid production for meeting the immediate mounting demands of the burgeoning human and livestock population without long-term perspective leads to unabated degradation of resource base, loss of biodiversity, decline in total factor productivity in the farming systems and environmental degradation. With shrinking per capita availability of land and water, coupled with pollution, the prevalent increasing demographic trend is putting further enormous strain on the already overburdened resource base.

While expressing deep concern on the deteriorating environment and some papers made references to the policy issues at the national, regional and local levels and called for reorientation in the policies. Some discussed the importance of land and water resources and suggested a long-term perspective planning for these resources to achieve the goal of economic prosperity and environmental security to meet various competing demands.

Section II deals with 'the Sustainable Agricultural Development and Food Security.' In all 15 papers are included in this section covering different aspects—sustainable livelihood, population growth and food security, food security and geo-spatial analysis, agribusiness, science and technology for food security, concept and techniques for measuring food security, sustainability of the agro-ecosystem etc. and have pointed out the deficiency in programme and policy and neglect of environmental sustainability, and called for a halt to the indiscriminate exploitation of natural resources. Most of the papers stressed the need for ensuring sustainable availability of food, sustaining the productivity of the natural resources, ensuring adequate household incomes, providing entitlement to food, introducting national food security, minimizing economic and environmental costs, diversification of farming system, promotion of rural employment, provision of eco-technology for fighting the fatigue of the green revolution etc.

Section III of the book deals with 'Weather and Climate' and contains five papers. These papers identify the problems of the areas needing priority attention for promoting conservation of climate and weather resources for sustainable agricultural productivity and food security. The issues dealt with in this part include monsoon droughts, drought and environmental security, drought management, climate change, ENSO and drought, also cover aspects of ecological security for sustainable agriculture. In the absolute terms agricultural productivity is directly related to prevailing climatic conditions and any aberration in climate and weather results in low food grain production followed by famines. The threat of climate change and its impact on environment and food security is one of the major environmental issues of the present time. Considerable uncertainty still surrounds the magnitude and spatial pattern of expected climate change and the consequences on crop yields. It is also suggested to take on the weather and climate on priority basis as the main research agenda for natural resource management in our country.

Section IV deals with the aspects of 'Geo-spatial Technology Application' which consists of five papers. The papers facuses on the use of remote sensing, geopgraphic information systems (GIS), globlal positioning systems (GPS) etc. in the natural resource assessment and expected to bring in the desired S & T inputs for sustainable resource management, and also suggested to have a comprehensive agricultural resources informatics using geomatics technology at district level. An urgent need to develop metadata and application of open GIS model for optimal utilization of agricultural resource bases of India is also stressed.

Section V of this book deals with aspects of 'Environmental Change and Impacts'. This part consists of five papers pertaining to global environmental change, environmental degradation and land use changes, impact of environmental change on ornamental and medicinal plants, implication of biotechnology on vegetable crops etc. The paper etitled the 'Global Environment : Crossing National Boundaries' indicated that the drive to make more money, to consume more and own more material goods is laudable from a national economic perspective but certainly undesirable in trying to ensure a holistic balance between society and environments such societies live in. One wonders whether the senseless accumulation of meaterial wealth is only measure for the quality of life and human satisfaction.

Section VI deals with the 'Socio-cultural Aspects' which consists of only two papers. Paper one on 'population growth, economic development and food security in India' analyses trend of population growth in the world with special foucus on India and concludes that the problem of food security is the result of rapid population growth and underdevelopment. Even developing economies like ours quality of life is offset by rapid population growth and poor purchasing capacity of a big section of our population. Paper second on 'Ecology & Encounters' has discussed in detail, how land use and land cover change during the different political regimes from the medieval period have changed and consequently induced changes in cropping pattern which leads to degradation of agro-ecosystem.

The papers included in this volume are multidisciplinary in character and cover a wide range of spectrum, seeking to provide a better understanding of the complex nature of problems related to resource conservation, sustainable development and food security. Some papers appear to be out of place, however if looked into in detail these are found to be quite relevant. All the papers are written by different experts with differing perception, conviction and different levels of experiences. A recurring idea running through all the papers is the need for the developing countries to adopt developmental policies and programmes that are socially relevant, environmentally compatible and economically viable. The papers have also drawn attention of politician, policy makers and policy implementers to be highly careful in choice and adoption of policy and use and ensure food security.

The book will serve as a reference material to the researchers, students, planners, decision-makers and others interested in judicious integrated management of the natural resources for sustainable development and food security. I want to express my hearty thanks to the Concept Publishing Company, New Delhi for undertaking the publication of this volume. The list of contributors is given at the fag end of the book.

SECTION-III

WEATHER AND CLIMATE

35

Links Between ENSO Phenomena and Drought Episodes in India

A Critical Examination

Tapeshwar Singh

Introduction

Drought is amongst a few devastating natural disasters in India. Although there is no universally acceptable definition of drought, the word is associated with prolonged and abnormal rainfall deficiency. It is a recurrent phenomenon. Occurrence of drought episode upsets the country's food security by seriously hitting agriculture, which still continues to be the mainstay of our farming communities, Indian agriculture is said to be gamble with monsoon. Drought occurs in both high as well as low rainfall area. It is a creeping phenomenon, making an accurate prediction of either its onset or end is a difficult task. The pattern of drought occurrence and its frequency and severity varies regionally. In our country drought episode is considered inevitable and an unavoidable part of "*Karma*".

Thus, interannual as well as spatial variations in the summer monsoon rain over India have a profound socio-economic impact. Indian summer monsoon contributes nearly 80 per cent of the annual rainwater potential. Failure of monsoon leads to drought, crop failure and in the more extreme cases, famine. Droughts in successive years or prolonging continuously for more than one or two years have a disastrous effect on food production and the national economy. The cause of drought is a complex question and it needs comprehensive research. Some researchers have attempted to link drought episode with ENSO phenomena. There is, however, a growing consensus among the international scientific community that ENSO phenomena have links with the extreme weather events, i.e. cyclones, floods, droughts etc. The challenging question before us is whether these droughts over different pacts of the world occur randomly or appear simultaneously with some large-scale atmospheric and/or oceanic phenomena? Several studies over the past decade or so have shown that interannual variation in many parts of the world and ENSO (El Nino/ Southern Oscillation) events in the tropics are generally related, for example, drought in Australia, Indonesia, India and parts of Africa and heavy rain and floods in the Pacific coast of South America (Nicholls 1991[1], Ropelewski and Harpert, 1987[2]). Prior to Sikka (1980[3]) many workers Lockyer and Lockyer (1902[4]), Walker (1923[5], 1924[6]), Bjerknes (1969[7]). Shukla & Misra (1977[8]), Khandekar 1979[9]) and others had also drawn attention on the links between global-scale oscillations in surface pressure and summer monsoon rain over India and surrounding regions. However, Sikka (1980) for the first time, related monsoon droughts over India with the occurrence of ENSO event. Since then several workers have attempted to study the links between Indian monsoon rainfall and ENSO/anti-ENSO events (Angell, 1981[10], Rasmusson and Carpenter 1983[11], Mooley and Parthasarathy, 1983[12], Parthasarathy and Pant, 1955[13], Bhalme *et al.*, 1980[14], Webster and Yang, 1992[15], Mohanty and Ramesh, 1993[16], Chattopadhyay and Bhatia, 1996[17], Sikka, 1998[18], Kane, 1995[19] and Victor *et al.*, 1995[20]). ENSO has been subject to major diagnostic and modelling studies in the past 25 years or so, particularly after the 1982 ENSO event. Although most of the studies suggest some links between ENSO (anti-ENSO) events and deficient summer monsoon rainfall over India. Different workers selected the ENSO years based on differing criteria, for example, Sea Surface Temperature Anomaly (SSTA) over eastern Pacific, Quinn *et al.* (1987[21]), Rasmusson and Carpenter (1982[22]), Southern Oscillation Index (SOI), Harpert and Ropelewski (1992[23]) and several other scientists have used atmospheric and oceanic parameters like disruption of fishery and marine bird life off the coasts of Peru and Equador. In light of this ambiguity in the definition of ENSO years serious differences prevail among the scientists about the listing of the ENSO years. Moreover, a confusion prevailed when ENSO event continues for two years, which year should be identified as ENSO year for its teleconnection study with summer monsoon rainfall. However, for the present study years defined by different workers like Deser and Wallace (1987[24]), Quinn *et al.* (1987), Shukla and Paolino (1983[25]), Rasmusson and Carpenter (1983), Sikka. (1998) and others are considered as ENSO year. In this study no attempt has been made to define ENSO years. Before a discussion on the links between ENSO phenomena and drought episodes in India, a brief account of ENSO is felt necessary. The El Nino/Southern Oscillation is one of the most pronounced Oscillations in the tropics. El Nino now more generally refers to as warming of the tropical Pacific basin that occurs roughly every three to seven years in association with a weakening of the trade winds. The flip side of El Niño, La Niña is characterized by stronger than normal trade winds and usually cold sea-surface temperature in the tropical Pacific. Both El Niño and La Niña are accompanied by swings in atmospheric pressure between the eastern and western Pacific known as the Southern Oscillation. These phenomena are collectively referred to as ENSO (El Niño/Southern Oscillation) Mcphaden (1999[26]). The general mechanism, underlying ENSO involves large-scale ocean-atmosphere interactions and equatorial ocean dynamics (Neelin *et al.*,1998[27]). But each El Nino and La Nina is unique in the combination of its strength, duration and pattern of development. Irregularity in the ENSO cycle can be seen in both instrumental record dating back to the middle of the last century and in proxy data (Such as late sediments, coral growth rings and ice cores) going back hundreds or even thousand of years (Rodbell *et al.*, 1999[28]). So in principle, it should not be surprising that an unusually strong El Niño occurs every so often (McPhaden, 1999). The two phenomena El Niño/Southern Oscillation are atmospheric and oceanic parts of an elegant climate fluctuations (Obasi, 1996[29]). Although it involves only one-fifth of the circumference of the planet but it transforms weather around the globe. According to historical record 600,000 people died in just one region of India from the epic droughts of the 1789-1793 El Niño (Supple, 1999[30]). The evolution of ENSO exhibits conspicuous irregularities from one cycle to another (Wang *et al.*, 1999[31]). The ENSO years listed in our study are 21 in number, i.e. 1877, 1884, 1891, 1896, 1899, 1902, 1905, 1911, 1918, 1925, 1928, 1941, 1951, 1965, 1969, 1972, 1977, 1982, 1987, 1992 and 1998. However, we have identified drought years on the basis of long time rainfall data series—annual, seasonal and homogeneous regions of India.

Objectives

1. To compute drought years for all India rainfall data series during the period 1871—1998 (128 years).
2. To assess the link between ENSO event and drought episodes on different spatial scales.
3. To examine the links between ENSO events on the simultaneous performance of annual, summer monsoon for All India and five homogeneous regions rainfall and to list non-ENSO years but drought as well as non-drought ENSO years.

Thus, the approach in our study is to examine the links between ENSO years and All-India drought episodes and summer monsoon season as well as different homogeneous regions. We have also attempted to discuss some peculiar features of ENSO Drought relation.

Methods

In the present study we have used Parasarathy *et al.* (1995[32]) data sets. This includes a listing of monthly, seasonal and annual rainfall data series for the period 1871—1994 for All India, 5 homogeneous regions and 29 meteorological sub-divisions. It is prepared on the basis of a fixed and well distributed network of 306 rain-gauge stations over India, each representing districts of different sizes by proper area weightage. Rainfall data since 1995 onward are provided by the Hydrology Division of Indian Meteorological Department (IMD), New Delhi. The contiguous area considered measures 2.88×106 sq. km which represents about 90 per cent of the country's total geographical area. The mean, standard deviation and coefficient of variation are also presented. All-India annual, monsoon season and Homogeneous Region's rainfall normal as well as percentage departure from the normal for all ENSO/Drought years are also computed. Different methods are used by investigators to define meteorological drought such as Ramdas Method (1950[33]), IMD Method (1971[34]), Drought

Area Indices Method, Monsoon Deficiency Index (Mooley and Parthasarathy 1982[35], Parthasarathy *et al.*, 1992[36]) and Standard Deviation Methods. In the present study standard deviation method is used to classify the performance of monsoon rains based on the departure from the S.D. (6). This criteria is used for All-India, all 5 Homogeneous Regions of India as well as summer monsoon rainfall. On the basis of this criteria we have determined different categories of drought on different spatial scales. The drought is considered to have occurred if the deficiency of rainfall is equal to or above 1.0 δ.

On the basis of this method we have defined various classes of droughts. The drought is classified as *mild* if the rainfall deficiency is between 1.0 to 1.24 δ, *moderate* between 1.25 δ to 1.49 δ, *severe* between 1.50 to 2.0 δ and *phenomenal* if the deficiency is more than 2.0 δ. Negative departure of seasonal rainfall from the normal by 25 per cent to 45 per cent is categorized as *phenomenal* drought and if the negative departure from the normal is by 50 per cent or more it is said to be *calamitous.* Droughts which have occurred during the period under study on (i) All-India basis, (ii) Monsoon season basis, and (iii) Homogeneous Regions of India basis have been listed in Tables 35.1 and 35.2.

Results and Discussion

All-India drought : In this section we discuss the occurrence of drought on All-India and summer monsoon season basis. Table 35.1 reveals that there have been 23 years when summer monsoon drought has occurred when the deficiency of rain has been 1.0.δ or more. And of these 17 years witnessed moderate drought, 14 years severe and only 4 years phenomenal, viz. 1877, 1899, 1951 and 1972 and all these years are ENSO years. It is also reflected from this table that the frequency of drought was more during the period 1899 or 1920, it decreased sharply during 1921 to 1960 and again increased during 1961 to 1990. The period 1921 to 1964 (43 years) did not witness a single severe and phenomental drought despite 3 ENSO events of 1925, 1941 and 1951. We also notice from this table that there are 4 years since 1871 to present, that India as a whole suffered from monsoon rainfall deficiency ranging between 25 per cent to 49 per cent of the normal. There is no year on record on All-India scale

Table 35.1 : Indian Monsoon and Homogeneous Region Droughts (1871-98) *(Rainfall Departure in Percentage)*

S. No.	Drought Years	Standard Deviation(δ) Seasonal	All India Annual	All India India Region	North-West India Region	West Central India	Central North-East India Religion	North-Eaat India Religion	South Peninsular India Religion
1.	1873	-1.0	-11.15	-13.96	3.00	-10.80	-12.73	-17.85	-15.39
2.	1877*	-2.9	-29.11	-13.15	-61.50	-28.80	-38.68	-3.8	-9.60
3.	1899*	-2.6	.26.17	-25.57	-66.90	-42.92	-1.11	15.64	-34.94
4.	1901	-1.6	-15.29	-10.84	-46.45	-9.09	-18.52	-4.05	-9.40
5.	1904	-1.2	-11.97	-10.61	-36.55	-16.88	6.20	-6.33	-13.72
6,	1905*	-1.6	-15.93	-15.56	-45.29	-21.05	-8.39	9.35	-16.65
7.	1911*	-1.4	-13.59	-11.27	-55.67	-13.36	1.85	-0.23	-17.11
R.	1918*	-2.1	-23.63	-21.30	-64.31	-32.58	-13.00	26.35	-38.61
9.	1920	-1.7	-15.63	-15.46	-21.96	-29.14	-12.47	-5.76	-8.37
10.	1928*	-0.9	-9.89	-4.95	-12.61	-7.79	-16.05	-1.45	-12.41
11	1941*	-1.4	-14.55	-12.15	-12.69	-26.93	-8.96	0.77	9.28
12.	1951*	-2.0	-13.29	-10.80	-43.16	-12.87	-14.90	4.92	-3.37
13.	1965*	-1.5	-16.78	-20.00	-25.86	-19.38	-21.01	-0.15	-12.30
14.	1966	-1.4	-13.20	-11.03	-16.39	-19.53	-22.38	8.52	-0.36
15,	1968	-1.1	-11.47	-11.82	-30.16	-15.85	-12.70	8.47	-2.74
16.	1972*	-2.3	-23.40	-22.18	-36.59	-26.70	-19.30	-14.92	-18.30
17,	1974	-1.2	-12.24	-10.73	-38.73	-18.96	-16.01	19.48	9.46
18.	1979	-1.9	-16.96	-11.74	-3.08	-20.18	-34.10	-5.08	-3.58
19.	1982*	-1.3	-13.73	-10.18	-31.27	-10.59	-10.01	-13.83	-11.97
20.	1985	-1.0	-10.86	-7.62	-30.37	-13.89	1.76	-7.78	-9.65
21.	1986	-1.3	-12.83	-8.80	-30.41	-10.86	-7.76	-10.87	-10.92
22.	1987*	-1.8	-18.20	-11.48	-59.76	-20.85	-11.53	-12.00	-18.30
23.	1991	-1.0	-7.96	-7.80	-18.93	-10.44	-9.55	-7.40	16.38
Mean --		--	852.40	1090.30	490.00	933.20	1002.40	1419.20	659.40
Per Ann-		--	78.20	100.30	89.90	86.30	83.30	68.70	57.00
STD. --		--	84.70	103.30	132.40	125.90	112.80	121.30	98.30
COV. --		--	9.90	9.30	27.00	13.05	11.02	8.50	14.90

* ENSO events. Limit of 2.0 δ or 25% deficiency or more is phenomenal drought on All-India basis. Years 1918 and 1972 are approaching the limit marginally.

when the deficienty was more than 50 per cent. We also notice that during this period at least 21 ENSO events happened. If we compare the drought years over All-India basis with each of the 5 Homogeneous regions basis we found that hardly 50 per cent years are common to each other, even for moderate or severe drought events on All-India scale, some of the regions do not experience drought of the same magnitude in many of those years. In terms of mild and above classes it is almost similar but differs significantly in terms of moderate and above and severe and above classes. It shows that there is much variability on an interannual basis in the seasonal rainfall on homogeneous regions basis than on All-India basis. The West Central India Region and South Peninsular India Region have much less number of droughts under moderate and above classes than All-India basis. Phenomenal drought over North-West India Region occurred in 1987 after 1918 (interval of 69 years), over West-Central India Region in 1972 after 1941 (interval of 31 year), over Central-North India Region in 1966 after 1877 (interval of 89 years), over North-East India Region in 1992 after 1896 (interval of 96 years) and over South Peninsular India Region in 1952 after 1918 (interval of 34 years). These deviations raise questions to investigate the causes of these variations. It must be noted, that the standard deviation as well as normal rainfall deficiency over different regions differ considerably. Thus, despite ENSO events probability of occurrence of phenomenal drought on All-India basis is very low, i.e. once in 30 years or so and the phenomenal drought of 1972 occurred after a lapse of 54 years. These figures further point to fairly high stability in the India's summer monsoon rainfall. However, there is no uniform cyclic trend of occurrence of drought on All-India and regional scales. Further we do not find any clustering of drought except on 2 occasions of two successive years—1904 & 1905 and 1965 &1966 only, and one occasion of 3 continuous drought years 1985,1986 and 1987.

Drought over Homogeneous Regions : We have shown in Table 35.2 the rainfall deficient years for each Homogeneous Region based on standard deviation from the normal. Percentage departure from the normal is given in Table 35.1.

Table 35.2 shows that for the individual homogeneous region 18 to 22 drought years have occurred under mild or above, 8 to 19 under moderate or above, 4 to 11 under severe or above and 3 to 7 under

Table 35.2 : Incidence of Droughts of different categories over Homogeneous Regions of India.

	Mild & Above $\geq 1.0\delta$	*Moderate & above* $\geq 1.25\ \delta$	*Severe & above* $\geq 1.50\ \delta$	*Phenomenal* $\geq 2.0\ \delta$
North West India Region	22	13	8	6
West Central India Region	20	10	10	7
Central North East India Region	20	15	6	3
North East India Region	22	19	11	4
South Peninsular India Region	I8	8	4	3

phenomenal drought. It shows that there is a considerable inter-regional variability with regard to different classes of drought. The phenomenal droughts are except North-west India and West-central India regions where 7 and 6 events respectively have occurred. It is important to note that even an area of high rainfall region of India such as North-east India Region is not ammune from such event, may these areas be rarely suffering from drought. It can also be noticed that among the common years of All-India scale severe to higher intensity drought and regional scale drought of the same magnitude, almost all years except 1920, 1974 and 1979 are ENSO years. This confirm an important feature of India's summer monsoon season drought, i.e. ENSO is linked to the occurrence of severe and phenomenal drought over All-India scale as well as on regional scale in North-west India region and West Central India Region. There are years when drought conditions prevailed over different regions while did not happen so over a contiguous homogeneous region or on All-India basis simultaneously, even during the ENSO years. It points to the fact that planetary scale feature like ENSO could be one driving forces of drought episodes in India but not exclusive. The interaction between planetary and regional scale features may result in chaotic nature of simultaneous occurrence of drought on All-India basis *versus* homogeneous region basis. The probability of occurrence of drought over entire areal extent of India simultaneously is almost negligible. Even in the worst drought years there could be meteorological sub-divisions where the rainfall would be normal or even excess. Thus, the scale of deficient or excess rainfall is not uniform on different spatial scales.

Non-ENSO years followed by monsoon droughts

It is found that of these 23 drought years, 11 years (1873, 1901, 1904, 1920, 1966, 1968, 1974, 1979, 1985, 1986 and 1991) are not associated with ENSO episodes. There is a secular variability in the ENSO-Monsoon relationship. This is a complex question and a challenge to climate science. It is expected that several questions relating to ENSO-Monsoon connections on interannual and decadal scale will be better answered through data and modelling studies during CLIVAR and ICRP in the next decade or so (Sikka, 1998). This indicates that the existence of an ENSO episode is not sure indication of impending monsoonal droughts. Thus, Indian monsoon performance may be associated with many other atmospheric, oceanic and terrestrial conditions. Therefore, ENSO is not a single condition for the drought occurrence in our country.

ENSO but not Drought

The years 1880, 1891, 1896, 1902, 1925, 1969,1983,1992 and 1998 were ENSO years but were not associated with incidence of drought event in India. During these years monsoon rains were either on positive side of the normal or did not show any significant departure from the normal. Fo[illegible]mple, despite *super ENSO* of 1998 this year India experienced mo[illegible]han 26 per cent rainfall from the normal and monsoon was considered to be excellently good. This poses a challenging question about ENSO-Monsoon relations, which we have to answer.

Conclusion

In the light of the above discussions we find that drought occurrence is frequent but has no cyclic trend. Evolution of ENSO also exhibits conspicuous irregularities from one cycle to another. The existence of an ENSO phenomenon is not a sure indication of impending monsoonal droughts since it has not consistently performed for the prediction of a monsoonal drought. The incidence of various classes of drought might be associated with other oceanic, atmospheric and terrestrial conditions, because atmosphere has a lot of chaotic noise.

ENSO is just one factor. However, ENSO is sure sign of non-occurrence of excess rainfall. Almost all ENSO years are associated with the deficient rainfall. Moreover, all 4 phenomenal droughts have links with ENSO events. ENSO events have a significant modulating effect on Indian monsoon. The result of this study by no means is perfect but certainly gives a much better indication of the climatic condition that is likely to prevail following ENSO strike. In the next millennium further detail investigations on a much larger database of ENSO-drought relations is felt necessary for objectively forecasting such happenings.

Acknowledgements

The author is grateful to the Director General, IMD, New Delhi and Dr. D.R. Sikka, Chairman MONTCIM, Department of Science and Technology (Government of India) and Dr. Shunil K. Dash, Principal Scientific Officer, Centre for Atmospheric Sciences, Indian Institute of Technology, New Delhi for constructive comments and suggestions.

NOTES

1. Nicholls, N. (1991) *Teleconnection and Heath-Teleconnection Linking worldwide climate anomalies.* (Cambridge University Press), 493-503.
2. Ropelewski, C.F. and Harpert, M.S. (1987, Global and regional scale precipitation patterns associated with theNino/Southern Oscillation. Mon. Weath. Rev., 115:1606-1626.
3. Sikka, D.R. (1980) Some aspects of large scale fluctuations of summer monsoon rainfall over India in relation to fluctuations in the planetary and regional scale circulation parameters. *Earth and Planetary Sci.,* 89: 179-95.
4. Lockyer, N. and Lockyer, W.J.S. (1902) On the similarity of the short-period pressure variation over large areas. *Proc. Roy. Soc.* London, 73: 457-70.
5. Walker, G.T. (1923) Correlation in seasonal variations of weather. *VIII Mem.* India Meteor Dept., 24: 75-131.
6. Walker, G.T. 1924. Correlation in seasonal variations of weather. *IX Mem.* India Meteor. Dept. 24 : 333-45.
7. Bjerknes, J. (1969) Atmospheric teleconnections from the equatorial Pacific. *Mon. Weath. Rev.,* 97:163-72.
8. Shukla, J. and Misra, B.M. (1977) Relationship between sea surface temperature and wind speed over the central Arabian sea and monsoon rainfall over India. *Mon. Weath Rev.* 105:993-1002.
9. Khandekar, M.L. (1979) Climate Teleconnections from the Equatorial Pacific to the Indian monsoon: analysis and implications. *Arch. Metero. Geophys.* Bioklim. A28 :159-68.

10. Angell, J.K. (1981) Comparison of variations in atmospheric quantities with sea surface temperature variations in the Equatorial Pacific. *Mon. Weath. Rev.* 109: 230-43.
11. Rasmusson, E.N. and Carpenter, T.H. (1983) The relationship between Eastern Equatorial Pacific sea surface temperature and rainfall over India and Sri Lanka. *Mon. Weath.Rev.* 111: 517-28.
12. Mooley, D.A. and Parthasarathy, B. (1983) Indian summer monsoon and El Nino. *PAGEOPH.* 121:339-52.
13. Parthasarathy, B. and Pant, G.B. (1985) Some features of a long homogeneous series of summer monsoon rainfall and the Southern Oscillation. *J. Clim.,* 5 : 369-78.
14. Bhalme, H.N., Sikder, A.B. and Jadhav, S.K. (1990) Relationship between planetary scale waves. Indian monsoon rainfall and ENSO, *Mausam*, 41(2): 279-84.
15. Webster, P.J. and Yang, S. (1992) Monsoon and ENSO: Selectively interactive systems. *Quart. J.R. Met. Soc.,* 118 : 877-926.
16. Mohanty, U.C. and Ramesh, K.J. (1993) Characteristics of certain surface meteorological parameters in relation to the interannual variability of Indian summer monsoon. *Proc. Ind. Acad. Sci.,* (Earth and Planet Sci.), 102: 73-88.
17. Chattopadhyay, J. and Bhatia, R. (1996) Are-examination of ENSO/anti ENSO events and simultaneous performance of the Indian summer monsoon. *Mausam.* A7(I) : 59-66.
18. Sikka, D.R. (1998) *Monsoonal Droughts in India.* In press.
19. Kane, R.P. (1998) El Nino, Southern Oscillation, Equatorial Eastern Pacific sea surface temperatures and summer monsoon rainfall in India. *Mausam*, 49(1): 103-14.
20. Victor, U.S. *et al.* (1995) El Nino's effects on southwest monsoon rainfall in Andhra Pradesh, India. *Drought Network News*, 7(2) : 20-23.
21. Quinn, W.H. *et al.* (1978) Historical trends and statistics of the Southern Oscillation, El Nino and Indonesian droughts. *Fish. Bull.,* 76 : 663-78.
22. Rasmusson, E.M. and Carpenter, T.H. (1982) Variations in tropical sea surface temperature and surface wind fields associated with the Southern Oscillation/El Nino. *Mon. Weath. Rev.* 110 : 354-84.
23. Harpert, M.S. and Ropelewski, C.F. (1992) Surface temperature patterns associated with Southern Oscillation. *J. Clim*, 577-93.
24. Deser, C. and J.M. Wallace. (1987) El Nino events and their relation to the Southern oscillation. 192S-986, *J. Geophys. Res.,* 14:189-96.
25. Shukla, J. and Paolino, D.A. (1983) The southern oscillation and long-range forecasting of the summer monsoonal rainfall over India. *Mon. Weath. Rev.,* 111:1830-1837.
26. McPhaden, M.J. (1999) The Child Prodigy of 1997-1998. *Nature*, 390 : 559-62.
27. Neelin, J.D. *et al.* (1998) *J. Geophys. Res.,* 103:14261-14290.
28. Rodbell, D.T. *et al.* (1999) *Science.*, 283: 950-54.
29. Obasi, G.P.O. (1996) Climate, *Climate change, variability and productivity.* Rajiv Gandhi Institute for Contemporary Study, Paper No. 26.
30. Supple, C. (1999) Nature's Vicious Cycle. *National Geographic*, 105 : 73-95.
31. Wang, B. *et al.* (1999) Stochastic Dynamics of El Nino-Southern Oscillation. *WANGETAL*, Jan. 1999:5-23.

32. Parthasarthy, B., Munot, A.M. and Kothawale, D.R. (1995) *Monthly and seasonal rainfall series for All-India, homogeneous regions and meteorological sub-divisions,* 1871-1994. IMD, Pune.
33. Ramdas, L.A. (1950) Rainfall and agriculture. *'Ind. J. Met. Geonhvs'*, 1: 262.
34. I.M.D. (1971) *Rainfall and droughts in India.* IMD, Pune.
35. Mooley, DA. and Parthasarthy, B. (1982) Fluctuations in the deficiency of the summer monsoon over India and their effect on economy. *Arch. Met. Geophys.* Bioklim Ser, B., 30: 383-98.
36. Parthasarathy, B., Rupa Kumar and Kothawale, D.R. (1992) Indian summer monsoon rainfall indices 1871-1990. *Met. Mag.* 126:174-186.

36

Climate Variability and Change and their Impact on Land Degradation in Dryland Regions of India

Y.S. Ramakrishna, G.G.S.N. Rao, G. Nageswara Rao and *P. Vijaya Kumar*

Overview of Land Degradation in India

Land degradation is a serious environmental problem in the dryland regions of India, which cover about 203 million ha area (62 per cent of the total geographical area) and consist of arid, semi-arid and dry sub-humid zones. Land degradation—the deterioration or loss of the productive capacity of land for present and future use—occurs mainly because of various forms of soil erosion (by water and wind) and of chemical and physical deterioration. Latest estimates (Table 36.1) given by NBSS & LUP (Sehgal and Abrol, 1994) show that about 57 per cent of land (187.8 million ha) in the country is exposed to various degradation factors; of which a dominant fraction (45 per cent) is due to water erosion (148.9 million ha) and the rest about 12 per cent is due to wind erosion (13.5 million ha), chemical deterioration (13.8 million ha) and physical deterioration (11.5 million ha). Land degradation/soil erosion results from various factors, including climatic factors and management practices, such as overgrazing or unsuitable cultivation practices, which leave the land vulnerable during rimes of erosive rainfall or windstorms. Climatic factors include frequent floods, rains of high intensity, blowing of high velocity winds and drought conditions in dry regions.

Table 36.1. Soil erosion in India

Degradation factors	*Area in million ha (Per cent of the total geographical area)*
Water erosion	148.9 (45.0%)
Wind erosion	13.5 (4.1%)
Chemical deterioration	13.8 (4.2%)
Physical deterioration	11.6 (3.7%)
Total	187.8 (57.0%)

Source : Sehgal and Abrol,1994.

The main factors contributing to water erosion, summarized by the Universal Soil Loss Equation (Wischmeier and Smith, 1978) are rainfall erosivity, determined by the energy, intensity and duration of rainfall; the erodibility of the soil; topography; vegetative cover and farmers' management practices. Except for farmers' management practices and vegetative cover, these are largely naturally determined factors, although human activities do influence soil texture and organic matter content to an important extent. Water erosion accounts for 87 per cent of the total degraded land in India. The states of Gujarat, Maharashtra, Andhra Pradesh, Tamil Nadu and Kerala are the worst sufferers of erosion. It results in a loss of topsoil in 130.5 million ha and terrain deformation in 16.4 million ha. Based on the analysis of existing soil loss data, Narayana and Ram Babu (1983) estimated the average rate of soil erosion at 16.35 t/ha/yr, yielding a total soil loss of 5.3 billion tonnes annually in the country. The annual erosion rate ranges from 5 t/ha for dense forests, snow-clad cold deserts and arid regions of western Rajasthan to more than 80 t/ha in the Shiwalik hills (Gurmel Singh *et al.*, 1990).

Wind erosion is determined largely by wind velocity, soil erodibility, topography, vegetative cover and farmer's management practices (particularly with regard to windbreaks). It is a serious problem in the arid and semi-arid regions including the states of Rajasthan, Haryana, Gujarat and Punjab. Wind erosion is also prevalent in the coastal areas where sandy soils predominate and in the cold desert regions of Leh in extreme north-western India. About 4.1 per cent of the total geographical area in the country is affected by wind erosion (Table 36.1). The threshold velocity for initiating wind erosion is estimated to be around 10 km/hr. The arid and semi-arid regions of the north-west cover 28,600 square kilometres, of which the sand dunes and sandy plains of western Rajasthan, Haryana, Punjab and Gujarat account for 66 per cent (Gupta, 1990). Severe wind erosion is observed mostly in the extreme western sectors of the country.

Climate Variability and Change and their Impact on Land Degradation

Rainfall is the leading soil forming factor that affects the type and predominant properties of soils specific to a region (Katyal and Reddy, 1997). The limited precipitation and rapid evaporation, characteristic of arid regions limit the depth of penetration of rain and as a result soluble substances accumulate at relatively shallow depths. Soil erosion/land degradation occurs both incrementally as a result of many small rainfall events, and more dramatically, as a result of large but relatively rare storms. Dramatic erosion events can produce large gullies and create flooding and property damage. By contrast, erosion due to small, common events may appear insignificant on the field but its cumulative impact can be equally severe. Land degradation is enhanced during periods of drought consequent upon the drying out of topsoil and effective loss of soil structure and aggregation. In this condition, topsoil is easily blow away as a result of the erodibility of the soil and the erosive nature of wind and rainfall.

Climate variability

Indian climate has a unique distinction because of the pervasive influence of the monsoons. The summer south-west monsoon influences the climate of the region from June to September, and the winter north-east monsoon controls the climate from November to February. The monsoon brings most of the region's precipitation and is the most critical climatic factor in the provision of water resource for rainfed and irrigated agriculture. The El Niño-Southern Oscillation (ENSO) phenomenon, which is geographically very extensive, has an especially important influence on the interannual variability of climate. The dryland regions in the country are highly prone to climate variability, as in these regions drought is a regular part of the natural cycles affecting productivity and leading to land degradation and desertification. Drought disasters are more frequent in India during the years following warm ENSO (El Niño) events, which are characterized by warm sea surface temperatures along the South American (Peru) coast around December. Similarly, the opposite cold ENSO (La Niña) events are associated with floods.

When several decades were combined to provide an understanding of the decadal scale variability of drought occurrence in India, it is found that the 1891-1920 and 1961-80 periods witnessed frequent droughts while few droughts occurred during 1930-1960 and 1980-2000. This suggests some kind of low frequency oscillation of the monsoon system on the decadal scale (Das, 2000). Gregory (1994) observed that over arid and semi-arid regions of India there have been periods in the last decades of the nineteenth century and first two decades of twentieth century when drought frequencies have been markedly higher than in recent years, although individual recent droughts may have been intense ones in some regions.

Low rainfall in arid and semi-arid regions dictates the formation of shallow soils, which are poor in organic matter and nutrients. Coupled with the conversion of grazing lands to farming, the arid region of India is undergoing an aggravation of desertification through erosion of lands and Aeolian shifting of soil particles. In some locations there has been a rise of water table with simultaneous increase of salinity and deterioration of soil regime. The variation of the aridity index line, computed by Penman method, over arid regions of India reveals a possible spread of arid conditions in the south-east direction (Jain, 1986). Studies (Ramakrishna, 1988) indicate that even on a decadal basis there had been shifts in the areas experiencing extreme aridity conditions (mean moisture index of 80.0) over the western Rajasthan region (Fig. 36.1). Studies on spectral variability of aridity over north India (Singh *et al.*, 1992a, b) indicate considerable expansion and contraction in the areas experiencing arid conditions in different years of the study period, influenced by the year-to-year variability in rainfall pattern. Analysis of the relationship between frequency of dust storm activity and monsoon rainfall at Jodhpur in the Indian arid zone (Fig. 36.2) shows that the years following low monsoon activity experience increased dust storm activity as the limited rainfall during the previous monsoon season leads to decreased

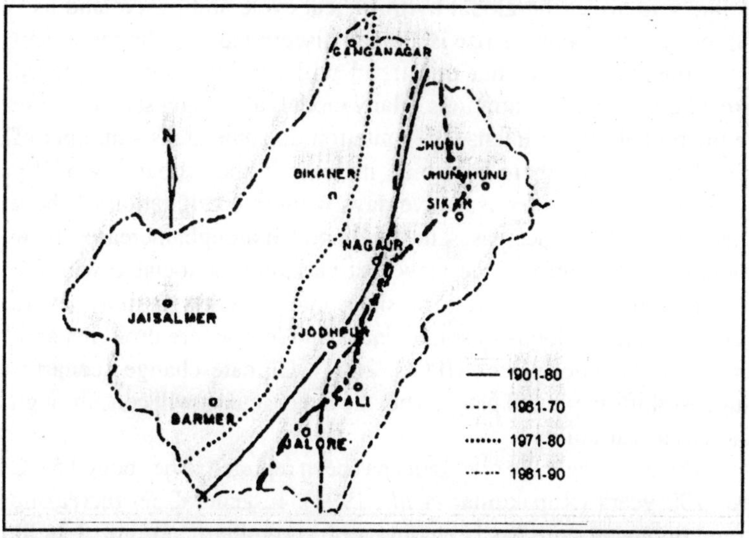

Fig. 36.1 : Shifts in Aridity Index (Line of 80%) over Western Rajasthan.

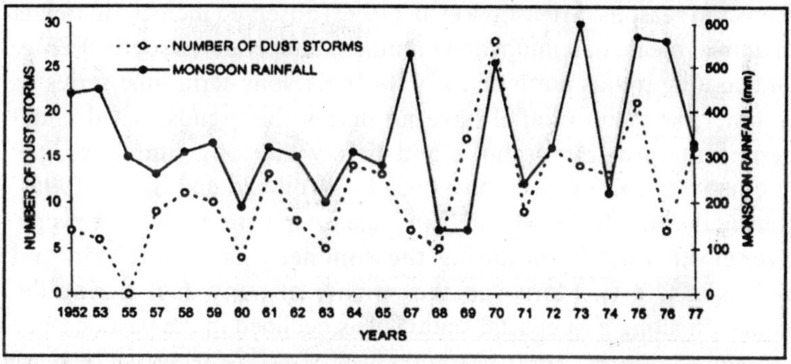

Fig. 36.2 : Relationship between frequency of dust storms and monsoon rainfall.

vegetation growth and consequent overgrazing and trampling by livestock exposing bare and loosened topsoil for wind erosion.

Climate change

In addition to the natural climatic variability, human-induced climate change also affect the climatic conditions of drylands, and hence their degradation. Global warming under the increased atmospheric CO_2

is likely to affect the global hydrological cycle and hence land use. Globally, a temperature rise is already discernible and climate model experiments suggest that this trend will continue with increased greenhouse gas concentrations. Many models also suggest an increase in the probability of intense precipitation. In some areas a number of simulations show there is also an increase in the probability of dry days and dry spells (consecutive days without precipitation). Where mean precipitation decreases, the likelihood of drought increases. New model results reinforce the view that variability associated with the enhanced hydrological circle translates into prospects for more severe droughts and/or floods in some places and less severe droughts and/ or floods in other places (IPCC, 2001). Climate change scenarios suggest shifts in rainfall patterns but the major impact will cone through increased variability.

The warming trend over India has been reported to be about 0.57°C per 100 years (Rupakumar *et al.*, 1994). In general, an increasing trend in temperature has been observed in southern and central India in recent decades in all seasons and over all of India in the post-monsoon seasons. Srivastava et al. (1992) observed increase in trends of annual mean, maximum and minimum temperatures south of 23°N and cooling trends north of 23°N. In India, long-term time series of summer monsoon rainfall have no discernible trends, but decadal departures are found above and below the long time averages alternatively for 3 consecutive decades (Kothyari and Singh, 1996). Recent decades have exhibited an increase in extreme rainfall events over north-west India during the summer monsoon (Singh and Sontakke, 2001). Moreover, the number of rainy days during the monsoon along east coastal stations has declined in the past decade.

Future Climate Change

Lal *et al.* (2001) presented the future climate change scenarios for the Indian subcontinent as simulated in the coupled atmosphere-ocean General Circulation Model (A-O GCM). Their study suggested an annual mean area-averaged surface warming over the Indian subcontinent to range between 3.53 and 5.55°C (Table 36.2) over the region during 2080s. During winter, India may experience between 5 and 25 per cent decline in rainfall. The decline in wintertime rainfall over India is likely to be significant and may lead to droughts during the dry summer months. Only a 10 to 15 per cent increase is projected in area-averaged summer monsoon rainfall over the Indian subcontinent. Their study also revealed more intense rainfall spells in a warmer atmosphere, thus increasing the probability of extreme rainfall events.

Table 36.2 : Climate change projection for Indian subcontinent

Year	*Season*	*Temperature change (°C)*		*Precipitation change (%)*	
		Lowest	*Highest*	*Lowest*	*Highest*
	Summer	0.87	1.17	1.81	5.10
2020s	Winter	1.08	1.54	-1.95	4.36
	Annual	1.00	1.41	2.16	5.97
	Summer	1.81	2.37	7.18	10.52
2050s	Winter	2.54	3.18	-9.22	3.82
	Annual	2.23	2.87	5.36	9.34
	Summer	2.91	4.62	10.10	15.18
2080s	Winter	4.14	6.31	-24.83	-4.50
	Annual	3.53	5.55	7.48	9.90

Source : Lal *et al.*, 2001.

Climate change; higher average temperature and changing precipitation patterns, can have three direct impacts on soil conditions. The higher temperatures cause higher decomposition rates of soil organic matter. Soil organic matter is important for ecosystem functioning. It is a source of nutrients, it improves aggregate stability, moisture storage capacity, and microbial activities. Generally, the chemical, physical and biological status of the soil increases with higher organic matter percentages (Sombroek *et al.*, 1993). Especially the decrease in aggregate stability may enhance water erosion. More floods will cause more water erosion, while more droughts will cause more wind erosion. All changes will require adaptation to the new conditions.

Reducing the Risks of Land Degradation

Land degradation results due to climate extremes, such as droughts or extreme wet seasons in combination with inappropriate grazing and resource management decisions. Climatic extremes and wrong decisions will lead to be the collapse of the pasture resource's ability to support animals, the loss of surface soil protection and the death of perennial vegetation. Hence, advance information on climate variability would be useful for the land managers to prevent/reduce the risk of land degradation. Overgrazing during the periods of rainfall deficiency will cause loss of desirable perennial plant species. Stock adjustment with seasonal forecasting could therefore prevent the accumulation of harmful grazing effects. Overgrazing in the first year of extreme rainfall deficiency causes resource damage and leads to rapid losses of potential productivity in subsequent years. In this situation, seasonal and annual forecasts for extreme drought periods will be useful. When extreme drought periods are greater than one year the length of drought is such that available vegetation is unable to supply adequate and surface cover for the extended period of rainfall deficiency. Therefore, forecasts of when drought duration is likely to exceed one year will be required. However, the current seasonal climate forecasts do not provide information on the magnitude of rainfall deficiency or duration. Hence, these forecasts need to be modified in the future to provide more useful warnings to prevent these land management problems.

Summary and Conclusions

There is a significant association between soil degradation, water management and climate variability/change. While frequent floods and high intensity rainfall events cause more water erosion, frequent and prolonged droughts cause wind erosion. Higher rainfall will probably intensify degradation, particularly in the dryland regions; which, without adequate responses, will magnify the problems of water pollution and clogged downstream watercourses, including reservoirs. Such changes could also result in the more intensive use of marginal soils to compensate for the loss of degraded land. Similarly, frequent droughts enhance soil erosion/land degradation because of the drying out of topsoil and effective loss of soil structure and aggregation. Climate change, resulting in higher average temperature and changing precipitation patterns, will further enhance the risk of land degradation, particularly in the dryland regions in the country. The risk of soil erosion/land degradation can be reduced through proper land management, particularly during the periods of climate extremes. This requires advance information on climate variability such as the occurrence of floods/droughts and the magnitude and period of rainfall deficiency etc.

REFERENCES

Das, H.P., (2000) Monitoring the incidence of large-scale drought in India. In: *Drought: A Global Assessment.* Vol. 1, Wilhite, Donald A. (Ed.)., Routledge, London and New York, pp. 181-95.

Gregory, S., (1994) The changing frequency of the drought in India, 1871-1985, *Drought Planning in India.* Printwell. Jaipur, India.

Gupta, J.P., (1990) Sand dunes and their stabilization. In *Technologies for wasteland development,* Abrol, I.P. and Narayana, V.V.D. (Eds.), New Delhi, ICAR Publication, p.253.

Gurmel Singh, Ram Babu, Narian, P., Bhusan, L.S., Abrol, Y.P., (1990) Soil erosion rates in India, *Jr. of Soil and Water Conservation,* 47 (1), 97-99.

IPCC, (2001) *Climate Change 2001: The Scientific* Basis. Contributiøn of Working Group I to the Third Assessment Report of the Intergovernmental Panel on Climate Change [Houghton, J.T., Y. Ding, D.J. Griggs, M. Noguer, P.J. van der Linden, X. Dai, K. Maskell and C.A. Johnson (Eds)] Cambridge University Press, Cambridge, United Kingdom and New York, NY, USA.

Jain, J.K., (1986) Combating desertification in developing countries, *Scientific Reviews on Arid Zone Research.*4, Scientific Publishers, Jodhpur, India.

Katyal, J.C. and Reddy, K.C.K. (1997) Plant nutrient supply needs: rainfed food crops. In *Plant Nutrient Needs, Supply Efficiency and Policy Issues: 2000-2015,* Kanwar, J.S. and Reddy, J.C. (eds.), National Academy of Agricultural Sciences, New Delhi, 91-113.

Kothyati, U.C., and Singh, V. P., (1996) Rainfall and temperature trends in India, *Hydrological Processes , 10,* 357-72.

Kripalani, R.H. and Kulkarni, A. (1996) Assessing the impacts of El Nino and non-El Nino-related droughts over India. *Drought Network News, 8,* 11-13.

Lal, M., Nozawa, T., Emori, S., Harasawa, H., Takshashi, K., Kimoto, M., Abe-Ouchi, A., Nakajima, T., Takemur, T. and Anumaguti, (2001) Future climate change: Implications for Indian summer, monsoon and its variability, *Current Science*, 81, 1196-1207.

Narayan, V.V.D. and Ram Babu, (1993) Estimation of soil erosion in India. In *Soil and Water Conservation Research in India,* Narayan, V.V.D. (Ed.), New Delhi, ICAR Publication.

Ramakrishna, Y.S., (1988) Climatic changes in relation to desertification in the Indian arid zone. In: *Desertification: Monitoring and control* (Ed. A.K.Tiwari), Agro-Botanical Publ., Bikaner, 27-35.

Rupakumar, K., Kumar, Krishna K., and Pant, G.B., (1994) Diurnal asymmetry of surface temperature trends over India, *Geophysical Research Letters*, 21, 677-80.

Singh, N., and Sontakke, N.A., (2001) Natural and anthropogenic environmental changes of the Indo-Gangetic Plains, India. Clim. *Change.*

Singh, N., Mulye, S.S. and Pant, G.B. (1992a) Some features of the arid area variations over India: 1971-1984, *Pure Appl. Geophy.*, 138, 135-150.

Singh, N., Pant, G.B. and Mulye, S.S. (1992b) Spatial variability of aridity over northern India, *Proc. Ind. Acad. Sci.* (Earth Planet. Sci.), 101, 3, 201-213.

Srivastava, H.N., Dewan, B.N., Dikshit, S.K., Prakash Rao, G.S., Singh, S.S., and Rao, K.R., (1992) Decadal trends in climate over India, *Mausam,* 43, 7-20.

Sehgal, J.L. and Abrol, Y.P. (1994) *Soil Degradation in India: Status and Impact.* Oxford and IBH Publishing Co. Pvt. Ltd., New Delhi, India, p.80.

Sombrock, W.G., Nachtegaele, F.O. and Hebel, A. (1993) Amounts, Dynamics and Sequestering of Carbon in Tropical and Subtropical Soils, *Ambio* 22(7): 417-426.

Wischmeier, W.H. and Smith, D.D. (1978) *Agricultural Handbook* 537. Washington, DC, U.S. Department of Agriculture.

37

Long-term Strategy for Drought Management

Kamta Prasad

Drought areas constitute a sizeable part of the country having 19 per cent of its total area and about 12 per cent of its population[1]. The areas go through all the suffering and deprivation which are characteristically associated with drought; they also form a significant part of the poverty concentration areas of the country. The traditional response of the government is to provide temporary relief in terms of cash, food and nutrition, drinking water, health care and fodder for cattle and to create short-term employment opportunities through relief works.

These measures, however, are mere palliatives, which fail to address the basics. The risk of loss due to drought remains unabated. Deficiencies in planning and execution of relief operations reduce their efficacy still further. Even, the quantum of relief provided is often quite meagre as compared to drought losses incurred.

According to a study made by the author, the quantum of relief provided to affected households during 1982 drought was almost negligible in Mahendergarh district of Haryana, being Rs. 400 to 549 per Block. Though higher in Udhampur district of Jammu & Kashmir and Purulia district of West Bengal, it was insignificant as compared to drought losses incurred, being 0.1 to 1 per cent of losses in Udhampur and one to two per cent in Purulia.[2]

Similar type of comments can be given on the usefulness of crop and cattle insurance which has been in operation in several states in the country since 1979. There are programme specific deficiencies like a very small proportion of farmers covered under the scheme. But the basic point is that the scheme is not designed for and is incapable of promoting sustainable development of drought-prone areas. It mitigates farmers losses after these have taken place; it does not prevent or reduce their occurrence. Nothing is done to reduce the environmental disadvantages associated with drought.

Thus, schemes like drought relief and crop and cattle insurance fail to contribute to the sustainable development of the drought prone areas so that their overall condition continues to remain the same. This is not to suggest that these measures be discarded altogether. Rather these should be strengthened further till better and long term alternatives become available. "Something is better than nothing." The on going communication revolution provides considerable opportunities for improving the relief administration. These should be availed of to the maximum extent. More attention be given to working out a better formula for determining the quantum of relief at the household level so that relief provided bears a reasonable semblance to the extent of the loss. Every effort be made to strengthen the agricultural extension outfit to reduce the quantum of damage. But the major thrust must lie in evolving and implementing term strategy for drought management so as to take the drought prone areas out of the drought syndrome altogether. A few suggestions in this respect are given below. These veer around two components, viz. water augmentation and more diversified economic structure.

Drought-Prone Area Programme (DPAP)

A programme for restoration of the ecological balance in drought-prone areas was started in 1972-73. Known as Drought-Prone Areas Programme (DPAP), it was originally started in 511 blocks of 70 districts in 13 states, but was extended subsequently to almost all the areas which deserved to be covered. The programme is continuing since then.

The programme is inherently sound and appropriate for drought-prone areas as can be seen from its principal components which are as follow : (1) Development and management of water resources; (2) Soil and moisture conservation; (3) Afforestation with special emphasis on social and farm forestry; (4) Development of pasture lands, and range management in conjunction with development of sheep husbandry; (5) Livestock development and dairy development; (6) Restructuring of cropping pattern and changes in agronomic practices: (7) Development of small-marginal farmers and agricultural labourers; and (8) Development of subsidiary occupations.

The programme was modified a few years ago by laying greater emphasis on watershed development.

But the impact of the programme has been quite limited, in fact almost insignificant principally because meagre resources have been provided resulting in their thin spread over wide areas. Lack of integrated approach has been another deficiency. Besides, inadequate attention has been paid to maintenance. This programme, therefore, can be an effective instrument for promoting sustainable development of drought-prone areas only if the quantum of investment per block is increased substantially and a more effective system of planning and monitoring is established. What is even more important is that it should be integrated with the on going programmes of development of water resources and other related sectors as discussed later.

Management of Rain Water, Watershed and Underground Water

The short-term impact of drought comprises of crop losses, scarcity of drinking water, loss of fodder for cattle, loss of employment for agricultural workers and adverse effect of all these on people's nutrition and health. In the case of disastrous or severe drought, death due to prolonged malnutrition is also possible. The long-term impact is to depress overall development of drought-prone areas below the level that would have prevailed had the areas been free from drought. The crucial effects are felt through agriculture which provides means of livelihood to bulk of population of these areas. Compared to drought free areas, agriculture in drought-prone areas has lower cropping intensity, less valuable cropping pattern, lower intake of fertiliser, lower per hectare crop yield and, therefore, lower per capita income from farming*.

* A comparison of drought and non-drought-prone areas of the three states of Haryana, J &K and West Bengal studied by the author showed that the drought-prone areas were dominated by relatively dry and less valuable crops like jowar, bajra, gram, barley and mustard, while the non-drought-prone areas were dominated by water-intesive and more valuable crops like paddy, wheat and cotton. Cropping intensity and crop yields were also higher for the non-drought-prone areas. The average per acre fertiliser consumption during 1980-83 was 9.28 and 24.86 kg respectively in drought and non-drought-prone districts of Haryana and 12.30 and 14.78 kg respectively in similar districts of West Bengal. (For details, see author's book *Drought, Disaster and Development,* Mittal Publication, New Delhi 1994).

The root cause the problem lies in inadequacy and uncertainty of rainfall. Hence, a basic remedy lies in provision of assured irrigation to the extent feasible. The people of the drought prone areas have a marked preference for irrigation as compared to drought relief as was revealed from the survey of drought prone areas in the three states conducted under my supervision.[3] It would, however, be advisable to prefer means of irrigation which have more favourable or less unfavourable effects on the environment so that development becomes sustainable. From this point of view, a high priority would go to rain water conservation as well as watershed management. Such an approach is preferable also on grounds of shorter gestation period, and greater amenability to local level decentralised planning. A major risk, however, lies in drying up of such water sources during periods of acute drought in all area. But such eventualities do not arise often.

According to our study covering 30 years period from 1951 to 1980, severe drought occurred only in two years in Harayana (Mahendergarh district), three years in Jammu & Kashmir (Udhampur district) and seven years in West Bengal (Purulia district). Disastrous drought was not experienced in any year in any of the three districts.[4]

Several experiments in both rain water harvesting and watershed management have been made in our coutry so that the technology is by and large well tested and familiar. Ministries of Agriculture and Rural Development and NABARD have launched national programmes of watershed development. A number of NGOs are also active in this field. There are several stories of both success as well failures from which appropriate lessons can be derived. But what is lacking and to which no attention is paid either by the government or by the NGOs in some estimate of the scope of these technologies in specific drought prone districts. Given the rain fall level and pattern and the topography, what proportion of the district's agricultural area can be irrigated and how much water would be provided for dirnking purpose? The picture could vary from district to district. In the absence of this information, it is not possible to judge the extent to which these technologies would provide a solution to the development of drought prone areas. I, therefore, make a strong plea for all the agencies concerned with this matter to launch an effort in this direction. Three steps are needed. First, a methodology based on studies in a few sample districts has to be eveolved. Second the district planning personal should be provided training in the appropriate methodology developed.

Third, the district planning machinery in drought prone areas has to be given the task along with funds.

Another option is to exploit the potential of ground-water resources. But the scope for this is drought-prone areas is limited on account of inadequate recharge through rains. Besides, over-exploitation is already resulting in progressive lowering ot the water table in many parts of the country—specially the drought-prone areas. Efficient exploitation of ground-water also depends on availability of electricity or other sources of energy which, however, are in short supply. Sustainability requires that preference be given to renewable sources of energy like hydro-power, bio-gas, windpower and solar energy. The growth of hydropower in our country has been quite slow. Alternative sources of energy have not developed so far. Solar energy seems to be ideally suited to environmental conditions of drought-prone areas, but solar or photovoltaic water pumping systems developed so far are highly uneconomical.[5] As regards windmills, the velocity of wind in many drought-prone areas is lower than the minimum velocity needed from the point of view of economic vaibility.[6] This leaves biogas as the only viable alternative."[7] Its scope, however, is also limited by availability of cattle and institutional constraints. Hence, for the time being, the main reliance has to be on electricity, the supply of which is limited. Moreover, it results in depletion of valuable coal reserves and create environmental problems which are not conducive for sustainable development.

Environmental Friendly Major Irrigation Works

Finally, there could be provision of canal irrigation through mass transfer of water from water abundant area. Large-scale irrigation projects often provide the best and most durable solution to the problem of drought as can be inferred from the experience of Punjab. Harayana and Rajasthan in our own country. Such projects, however, would imply construction of storage dams of different sizes, large medium and small. Transfer of water over long distances could require large dams. Construction of such dams is, however, opposed by a vocal and high profile group of environmental activists on the ground that these go against sustainable development. The objections relate mainly to : (i) submergence of forest land, (ii) effects on wild life and monuments, (iii) displacement of population, (iv) soil erosion and sedimentation of reservoir and canals and (v) waterlogging, soil salinity nd alkalinity asociated with irrigation. In some cases, seismic risks might also become important.

Most of the objections are based on the experience of the past when awareness of environmental concerns was no-existent or limited. The present senario is different. Environmental issues are coming to occupy the centre state in planning and implementation of developmental projects. Hence, it is rather unrealistic to proceed on the assumption of business as usual. Improvements are expected specially since the so called adverse environmental effects are not inevitable. Most of these can be and have sometimes been taken care of by a more careful planning at the initial stage itself. New norms and guidelines are in the process of evolution. Some useful suggestions have been made in the report of the National Commission for Integrated Water Resource Development, 1999. The World Commission oil Dams in its report submitted in November 2000 has also given a set of guidelines. I am happy to inform you that this Institute is currently engaged in the process of evolving a set of guidelines appropriate to India's socioeconomic and administrative framework for taking up large dams. The Institute's aim is to generate a national consensus on this issue within the next six to ten months. For this purpose, the Institute would be holding workshops in different parts of the country. On the basis of idea already developed by us. I am hopeful that this task, though very difficult, will be accomplished. National interest demands so specially since environmentally friendly major irigation projects provide the best strategy for dealing with the drought problem in our country.

Towards Greater Economic Diversification

The prospects for providing water through irrigation development as well as through rain water harvesting and watershed management in drought-prone areas are seem to be limited though a more precise view can emerge only after having a study of districtwise estimate of water potential of the type indicated earlier. It is, therefore, necessary to evolve a strategy for the areas where it is not possible to provide adequate water through any means.

It may be noted that the distress oil account of inadequacy of rain water becomes more pronounced in drought-prone areas because of excessive dependence of population on agriculture as a means of livelihood. And within agriculture, the emphasis is mainly on oil food grains. It is indeed an irony that the percentage of population depending upon agriculture is higher in drought-prone than in non-drought prone areas even though agriculture there is less profitable and more risky. The percentage of cultivators to workforce during 1980 in drought-prone and non-drought-prone districts was 40 and 36 respectively in Haryana, 72 and 54 in Jammu & Kashmir and 47 and 29 in West Bengal.[8] The greater the dependence on agriculture, the more vulnerable the drought-prone areas become to adverse effects of drought. Hence, for such areas, it would be necessary to lay emphasis on diversification of their economies. The greater the diversification, the greater the insulation from acute distress in years of drought.

Besides, there is need to diversify agriculture itself by changing the cropping pattern in favour of those crops which provide higher yield even under conditions of considerable soil moisture stress. Some varieties of a few crops suitable for drought-prone or dry areas have been evolved by agricultural scientists. These deserve to be encouraged.

What are the prospects for economic as well as agricultural diversification? The main reason for excessive dependence on agriculture and on food grains in particular in drought-prone areas lies in the absence of a reliable food security system. As a result, every village or a group of neighbouring villages try to produce enough food grains for its survival. The lesser the agricultural productivity, the greater the number of cultivators needed for raising the minimum quantum of food for survival. Farmers, therefore, would be willing to move away from agriculture and more particularly from raising food grains only if there is a reliable system of food security. This is an essential condition the significance of which needs to be realised at the earliest.

With regard to activities other than farming which should be developed in such areas, one may suggest forestry, horticulture, livestock and industries. Trees after they survive beyond their infancy, need less water than foodcrops. Livestocks need pastures and fodder crops which also need less water than foodcrops. A major breakthrough, however, would take place mainly through industrialisation. This is an activity whose requirement of water is the minimum. But our survey of drought-prone areas showed that the role of the industry as a source of income and employment in such areas has been extremely limited. Bold steps would, therefore, be needed for industrial development of drought-prone areas. The choice of industry, the scale of operation and technology, the development of infrastructural facilities, arrangement for supply of inputs, training of manpower, provision of markets, location of growth centres, and impair on environment are some of the aspects, on which detailed strategy would have to be worked out. Elaborate discussion of these lies beyond the scope of this paper and of this seminar. A separate seminar would be needed for that purpose. I would just conclude by submitting that the agencies dealing with handlooms, handicrafts, khadi and village industries, electronics etc. along with banking, financial, training and marketing institutions have to be mobilised to take up development of those drought prone areas having little prospect of water augmentation by any means on a priority basis. This task would be more productive than the never ending efforts and fund spent on relief operations.

Finally, a word of caution

In view of the prevailing diversity in the agro-climatic and socio-economic conditions of the drought-prone areas in India, it is not advisable to prescribe a uniform approach with respect to preferred set of activities, for the entire country. The approach has to be decentralised or area specific. The suggestions given in earlier sections may, therefore, be treated as broad indications rather than definite prescriptions.

REFERENCES

1. *Draft Fifth Five Year Plan,* 1974-79, Vol. II, p. 91.
2. Kamta Prasad - (i) "Report on Socio-Economic Conditions of Drought Prone Areas of Haryana, Jammu & Kashmir and West Bengal"; one volume on each, submitted to Central Water Commission, Government of India, New Delhi in December 1987. A revised and abridged version of these has been published as one volume entitled "*Drought Disaster and Development*" by Mittal Publications, New Delhi, 1994.
3. *Ibid.*
4. *Ibid.*

5. Kamta Prasad, "Report on Economic Viability and Organisational Requirements of Lift Irrigation Technologies for Small and Marginal Farmers in India" submitted to Council for Advancement of People's Action and Rural Technology (CAPART), New Delhi, Government of India, August 1990.
6. *Ibid.*
7. *Ibid.*
8. Kanita Prasad, *Drought Disaster and Development, op. cit.*

38

Role of Farm Extension in Drought Management

A Study of 1987 Drought in Haryana

R.K. Sharma

OBJECTIVES AND SCOPE

The main objective of this paper is to assess the role of farm extension in drought management during 1987 drought in Haryana. The study covers the following aspects of droughts and drought management in Haryana: (i) Pattern of droughts, (ii) The impact of droughts on crop areas, yield and output, (iii) Existing mechanism for monitoring of rainfall, preparation of contingency plans for crop stablisation and provision of extension services to farmers, and (iv) The performance of farm extension in minimizing the adverse impact of droughts particularly during the 1987-88 drought.

This paper is based on authors research study on this subject carried out in 1998 (Sharma, 1990). The study was based on a sample survey of 100 farmers in Mahendergarh and Bhiwani districts of Haryana and the regular 'Monitoring and Evaluation, (M&E)' surveys on the T&V extension carried out by the Department of Agriculture in the Kharif and Rabi seasons. The sample districts of Mahendergarh and Bhiwani were chosen for study because : (1) These districts are officially recognized as drought prone areas, and (2) These districts have the lowest proportion of gross cropped area under irrigation and therefore, suffer most during droughts.

DROUGHTS IN HARYANA

Since July rainfall is of critical importance for Kharif crops we have studied in detailed the behaviour of July rainfall at one location, namely Narnaul in the arid zone. The analysis is based on daily rainfall during the decade 1979-88 compiled by the Indian Meteorological Department. The data given in Table 38.1 clearly bring out the large fluctuations in July rainfall in respect of the number of rainy days, weekly distribution of rainfall and the intensity of precipitation.

In the absence of data on all the factors influencing drought, rainfall deficiency is used an indicator of the severity of droughts. The official definition of an agricultural drought is as follows:

(a) *Moderate drought*—when seasonal rainfall is between 50 to 75 per cent of the normal rainfall;

(b) *Severe drought* —when seasonal rainfall is less than 50 per cent of the normal rainfall.

Table 38.1 : Pattern of July rainfall at Narnaul in Mahendergarh District during 1979-1988

(Rainfall in mm)

	Particulars	*1-7 July*	*8-15 July*	*16-23 July*	*24-31 July*	*1-31 July*
1.	Number of rainy days					
	Average	1.8	2.6	2.2	2.3	8.3
	Range	0-4	1-4	0-6	0-4	2-14
2.	Rainfall					
	Average	12	53	36 .	45	146
	Range	0-50	7-123	0-153	0-146	36-260
	Per cent Share	8	36	25	31	100
3.	Highest rain in a day	35	51	60	60	60
4.	Number of wet spells of more than 1 day during 1979-88 (Number of Years)					
	2 days	3	.2	4	2	11
	3 days	–	3	–	2	5
	4 days	–	–	I	–	1
	Total	3	5	5	4	17
5.	Number of years					
	Without rain	5	–	2	1	–
	Below average rain	2	7	4	6	5
	Above average rain	3	3	4	3	5

On the basis of data on monsoon rainfall during 1871-84 for 29 meteorological sub-divisions, Parthasarthy has identified 22 droughts and 23 floods in Haryana meteorological sub-divisions during the same period (Parthasarthy *et al.*, 1987). The probability of the occurrence of droughts and floods in Haryana has been worked out as 0.193 and 0.202 respectively. In addition to these widespread droughts there are frequent prolonged local dry spells. For instance, during 1934-82, there were 23 dry spells of more than three months duration at Hissar (Bishnoi, 1988).

Table 38.2 shows the number of pronounced rainfall deficient years in various districts during July and August as well as for the monsoon season. During the decade 1976-86, as many as 7 districts had drought during 3 to 4 years in July. Only one district did not experience any drought in July during the decade. The occurrence of droughts was more during August. Two districts had drought in 5 years, 6 districts in 4 years and 4 districts in 3 years of the decade.

Table 38.2 : Number of Years of moderate and severe droughts in Haryana districts during the decade 1976-1986

Districts	*Total number of years of drought**			*Number of years of severe*** droughts*		
	July	*August*	*June-Sept.*	*July*	*August*	*June-Sept.*
I. Sub-humid Zone						
1. Ambala	0	4	2	0	1	—
II. Semi-arid Zone						
2. Kurukshetra	1	4	.2	1	2	—
3. Karnal	3	3	2	2	2	1
4. Sonepat	4	4	3	1	3	1
5. Jind	3	5	4	2	3	1
6. Gurgaon	3	4	3	1	2	1
7. Faridabad	3	4	3	1	2	1
III. Arid Zone						
8. Rohtak	4	4	3	2	2	1
9. Sirsa	2	5	4	2	4	I
10. Hissar	4	3	2	0	3	—
11. Bhiwani	4	3	4	3	3	1
12. Mahendergarh	4	3	3	3	2	2

* Rainfall deficiency of 25 per cent or more.

** Rainfall deficiency of 50 per cent or more.

The number of severe drought years were smaller, only 2 districts recorded maximum of 3 severe drought years in July. As is to be expected, the incidence of serve droughts was higher in August. One districts recorded 4 severe drought years and 4 districts had 3 years of severe drought in August during the decade 1976-86.

For the entire monsoon season (June to September) the number of drought years during the decade 1977-86 varied from 2 to 4 various districts (Table 38.2). Three districts recorded 4 drought years and five districts had 3 drought years. However, only one district recorded maximum of 2 severe drought years during the season. Thus, the incidence as well as the severity of the droughts varied significantly across districts.

Now about the two recent drought years, 1979-80 and 1987-88. During 1979-80 there was a severe summer drought in 3 districts in July and in all the districts during August. However, for the entire season only 4 districts recorded severe drought. On the contrary the 1987-88 drought covered all the districts. No districts recorded less then 50 per cent rainfall deficiency during the monsoon season. Indeed in July all the districts recorded more than 80 percent rainfall deficiency. In addition the year 1987-88 had pronounced rainfall deficiency even during the Rabi season.

The 1987 drought in Haryana covered all the districts and affected 6351 villages, 94 lakh rural population and 40 lakh cattle (GOI, 1989). The losses to crop output amounted to about Rs. 1100 crore (Rs. 700 crore in Kharif and Rs. 400 crore in Rabi). In addition milk output was badly affected. The state government had to undertake extensive measures to minimize the adverse effects of the drought. They include the following :

(a) Provision of extension services to help farmers in minimizing crop losses.
(b) Provision of farm input, specially seeds and energy (electricity and diesel) and subsidies and credit for the purchase of these inputs.
(c) Supply of drinking water, food grains and other essential goods like kerosene, provision of health care and nutrition for children and employment to the vulnerable sections.
(d) Relief for livestock—supply of fodder, drinking water, health care.

Indeed, a drought creates a situation of economic and social emergency and necessitates the mobilization of all official and voluntary agencies for minimizing the adverse impact of drought. The department of farm extension has a key role in the management of droughts.

As regards the impact of 1987 drought the Haryana Government estimates of expected loss in Kharif crop output was about Rs. 640 crore. About 39 per cent of the total loss was due to Bajra (including fodder), 32 per cent to Rice, 12 per cent to sugar cane and 8 per cent to cotton. The per cent reduction in output of major Kharif crop was as follows :

> Rice 56, Bajra 91, Sugar cane 38, Cotton 25, Jowar 44, Maize 61, Kharif pulses 69, Kharif Oilseeds 89.

The actual losses in output were much less than those estimated by the Government of Haryana. It may be noted that the estimates of loss prepared by Haryana Government for claiming drought relief were based on 'target' of production during 1987 and not the actual production during 1986-87. The estimates of the losses in area, yield and production of principal crops in Haryana during 1987 prepared by the Government of India are given in Table 38.3. It may be noted that the 1987 drought reduced the area under all Kharif crops except sugar cane and cotton which are sown before the start of monsoon. The reduction in area was quite substantial for Rice (26%), Bajra (38%), Maize (26%) and Kharif Oilseeds (42%). Further, there was reduction in yield of all Kharif crops, specially Jowar (59%), Bajra (49%), Maize (32%), Kharif Oilseeds (33%), Cotton (30%) and Sugar cane (32%). The production of major Kharif crops declined by more than 20 per cent. The reduction was substantial in the case of Rice (30%), Jowar (63%), Bajra (69%), Maize (50%), Oilseeds (61%).

The Rabi 1987 drought significantly reduced the area, yield and output of only gram by 67%, 51% and 84% respectively (Table 38.3). Its impact on wheat, which is predominantly irrigated, was negligible. On the other hand the area, yield and output of Rabi oilseeds increased by 19, 28 and 52 per cent respectively.

Let us now look at the impact of summer drought of 1987 on the area, yield and output of major Kharif crops in different districts of

Haryana. The analysis is based on provisional estimates prepared by the Government of Haryana. It may be noted that there are significant inter-district variations in the impact of drought in respect of area, yield and output of crops. As regards the impact of drought on crop area, it is evident that the area under major food grains (Rice, Bajra, Jowar and Maize) declined substantially in all districts. This is true even for heavily irrigated rice crop. At the same time the drought had no impact on area under cotton and sugar cane which are sown before the onset of monsoon. However, the yield of these crops as well as of Bajra, Maize and Jowar declined significantly in all the districts expect for Bajra in Gurgaon and Faridabad and Maize in Faridabad. The yield of rice declined in all districts except Karnal. Significantly, rice yield was marginally affected in three major rice growing districts of Karnal, Krukshetra and Ambala mainly because of the availability of assured irrigation from tube wells.

Table 38.3 : Impact of 1987 drought in Harvana—Per cent change in area, yield and output of major crops

Crops/Crop Groups	*Per cent Change in*		
	Area	*Yield*	*Output*
Kharif Crops			
Rice	-26	-05	-30
Jowar	-09	-59	-63
Bajra	-39	-49	-69
Maize	-26	-32	-50
Pulses	-03	-25	-27
Cotton	10	-30	-23
Sugar cane	15	-32	-22
Oilseeds	-42	-33	-61
Rabi Crops			
Wheat	-03	-01	-04
Gram	-67	-51	-84
Barley	-07	07	-01
Oilseeds	19	28	52

As regards crop output the per cent reduction in various districts varied from 14 to 67 in the case of Rice, 80 to 92 for Jowar, 13 to 84 for Bajra and 34 to 60 for Maize. Further, impact of drought on output of cotton and sugarcane was relatively smaller compared with other Kharif crops.

ROLE OF FARM EXTENSION

The effectiveness of extension in the management of abnormal situations, like drought, has to be judged in relation to the type of extension system followed, specially its manpower and other resources, tasks assigned to extension workers and support provided by research and other support services. In 1979 the extension system in Haryana was thoroughly reorganized and considerably strengthened in terms of manpower on the pattern of the T & V extension system evolved by the World Bank. The recognized system has the following features, namely (1) Unified extension system exclusively devoted to extension work and operating under a single line of command, (2) Continuous upgrading of the skills of extension workers, (3) Regular interaction between research and extension workers, (4) Regular fortnightly field visits of village extension workers to selected farmers and (5) Regular monitoring and evaluation of extension by the Department of Agriculture. Thus, the T & V extension system is an integrated system, jointly managed by the Department of Extension and the Haryana Agricultural University (HAU) and its scope is not confined to mere communication of technical advice to farmers by the extension workers.

The basic strategy for drought management in India has been outlined in details in the drought management manuals prepared by the Government of India from time to time. The latest manual "A Model Manual for Drought Management" was brought out by the Government of India during 1985. According to this manual drought is considered as "a creeping disaster" which develops gradually giving sufficient warning of its coverage etc. unlike floods which occur suddenly and offer little time for planning and preparation. The manual outlines the tasks of drought management in detail. These can be grouped into two broad categories, namely (a) measures for minimizing the impact of drought on crop output, and (b) relief measures for rural population and livestock.

As stated earlier, severe droughts create a situation of economic and social emergency and necessitate the mobilization of all official and voluntary agencies for handling the various tasks of drought management. Farm extension system is concerned with the first set of tasks which are to be carried out expeditiously.

The official manual on drought management also specifies the steps to be taken in minimizing the impact of drought. The first step is the formation of "Crop Weather Watch Group" at the State as well as district level. The function of this group, which is to consist of the officials of the departments of Agriculture and Meteorology and crop specialists of the Haryana Agricultural University, are as follows:

1. Keep a close watch on the behaviour of monsoon.
2. Chalk out contigency plans for crop pattern, input use and input supplies.
3. Convey these to the concerned official agencies and field staff.

Let us now look at the performance of the extension system in respect of various tasks, specially during 1987 drought.

Crop Weather Surveillance

In Haryana the task of crop weather surveillance is carried out jointly by the Department of Agriculture and the Haryana Agricultural University. The University has separate department of Agricultural Meteorology. This department obtains the latest IMD weather forecast for Haryana. The department of agriculture, in collaboration with agricultural scientists of the University issues a weekly one or two page bulletin on "Weather and Agriculture" in Hindi and English. This bulletin is also supplied to the regional stations of the All India Radio for broadcast. It may be noted that the weekly 'Weather and Agriculture' bulletin gives forecast about weather (rainfall, temperature, evaporation, humidity) as well as the advice of crop experts about important crop practices (field preparation, sowing etc.), and likely appearance of pest/insects and measures to combat these.

The recommendations given in the HAU pamphlets were meant to be broad guidelines. More detailed recommendations specific to local conditions were to be worked out at the monthly workshops and explained to the village extension workers at the fortnightly training sessions. The available evidence shows that the task of modification of production recommendations was not taken seriously even during the drought period. A perusal of the fortnightly message cards issued in three agricultural sub-divisions in one of the selected districts of the modification of recommendations to suit the soil moisture deficiency created by the drought. In fact the message cards of only one sub-division took note of the failure of rainfall. Further, most of the message cards were very brief (less than one page).

Contingency Crop Planning

The next step for drought management is the preparation of contingency crop plans for minimizing the adverse impact of drought on crop output. The contingency crop planning is essentially aimed at stabilization of crop output in the situation of soil moisture deficiency. Since the extent of expected soil moisture deficiency is not known precisely, several alternative plans have to be prepared and continuously modified to suit the changing weather conditions. The major components of the contingency crop planning strategy, as outlined in the 1985 drought manual, are (a) alternative crop pattern to suit the soil moisture availability, (b) compensatory cropping on area not sown due to drought during the sowing season, and (c) modification in input use to suit the available moisture.

The task of contingency crop planning is the joint responsibility of the Department of Extension of the Haryana Agricultural University (HAU) and the Extension Division of the Department of Agriculture. In the first week of August, HAU released 3000 copies of a 4-page printed pamphlet in Hindi 'Choice and Protection of Crops in the event of Drought or Late Rainfall'. A copy of the pamphlet, translated into English, is given as annexure. The pamphlet pointed out that (a) there was no sowing on unirrigated area, (b) less area has been sown under paddy and the area already sown was drying, (c) cotton and sugar cane crop were in poor condition, and (d) fodder crops were drying.

More detailed recommendations were made for tackling drought during the Rabi 1987-88 season. During September 1987, HAU released 5000 copies of pamphlets of 8 pages with a title 'Measures for Fighting Drought in Haryana'. This pamphlet contained recommendations about choice of crops and input use separately for : (a) unirrigated areas, (b) areas having good irrigation, and (c) areas having brackish water supply. A similar 8-page pamphlet "How to obtain good output of Rabi Crops in drought conditions" (5000 copies) was issued in early November 1987. In addition this bulletin gave suggestion for increasing productivity of irrigation. The recommendations given in the HAU pamphlet were meant to be broad guidelines. More detailed recommendations specific to local conditions were to be worked out at the monthly workshops and explained to the field extension workers during the fortnightly training sessions.

Modification of Production Recommendations

The T & V extension has built-in system for continuous modification of production recommendations. "The recommended practices must be continuously reviewed and altered to reflect changing production conditions (including, for example, unusual weather)". To be useful to farmers "production recommendations must develop in the process of continuous adjustments to changing economic and environmental situations". The production recommendations for different crops are prepared by the scientists of Haryana Agricultural University. These are finalized at seasonal workshops of extension workers and scientists of HAU before the start of Kharif and Rabi seasons. Recommendations are further refined at seasonal Zonal Workshops and modified at monthly workshops and fortnightly training sessions to take account of local field and production conditions.

The available evidence, however, shows that even under normal weather conditions there is very little indication about modifications of production recommendations to suit local conditions. The recommendations listed in the fortnightly 'message cards' are largely reproduced from the seasonal booklets on packages of recommendations issued by HAU. A perusal of the fortnightly message cards issued in three agricultural sub-divisions of one of the selected district during the period 13 July to 9 September, 1987, do not give any indication about the modification of production recommendations to tackle the problems arising out of soil moisture deficiency created by the drought.

Communication of Production Recommendations

The success of the extension system crucially depends upon the regularity and effectiveness of the Village Extension Workers

(VEW's) fortnightly visits to the selected contact farmers. For the reasons mentioned earlier, it was not possible to know from the farmers the exact number of VEW visits different fortnights of 1987 Kharif season. Instead, they were asked whether they received any advice or help from the VEW's during 1987 Kharif season. Significantly, majority of the contact farmers reported some interaction with the VEW's during the Kharif season. This is, however, not an indicator of the regularity of the VEW visits. In fact, the special drought subsidy on inputs resulted in more than usual interaction between VEW's and farmers.

The M & E reports provide a clearer picture about the regularity of VEW visits. The report for Kharif 1987 shows that only 69 per cent of sample contact farmers reported stipulated two VEW visits during the reference months of M & E survey. Further, 16 per cent contact farmers reported no visit during the month. The report is that the frequency of VEW visits during the drought period did not show any improvement compared with 1986-87 or earlier years. Thus, the VEW did not make any efforts to improve their performance even during the drought. It may be noted that M & E surveys cover about 2000 farmers in 950 villages in the state.

To sum-up, Haryana has a well-designed T & V extension system. However it is functioning much below expectations. We have shown that the role of extension during 1987 drought was no better compared with normal years. There is need to improve the system so that it can cope with the additional work required during abnormal weather specially in arid areas in the State.

REFERENCES

Benor, Daniel and Micheal Baxter, (1984) *Training and Visits Extension,* The World Bank, Washington.

Beran, Max, (1989) 'Drought: Processes and Future' in International Symposium on Groundwater Resources Management in Drought Prone Areas, 27 Nov. —1 Dec., 1989, New Delhi, Ministry of Water Resources, New Delhi.

Bishnoi, C.P. '*Agro-climatic Surveys of Haryana*', Haryana Agricultural University.

Government of Haryana, 'Statistical Abstract of Haryana' (Annual).

Government of Haryana, 'Production Programme for Rabi Crops, 1988-89,' Department of Agriculture, Chandigarh (Issued every year).

Government of Haryana, 'Production Programme for Kharif Crops, Department of Agriculture, Chandigarh (Issued every year).

Government of Haryana, (1987) 'Memorandum to Government of India on Drought Relief Operations, 1987', Revenue Department, August, 1987.

Government of Haryana, (1988), 'Monitoring-cum-Evaluation Study on functioning of Reorganised Agricultural Extension System in Haryana -Kharif 1987-88', Department of Agriculture, Chandigarh (Similar reports are brought out for Kharif and Rabi seasons every year).

Government of Haryana, (1989), Evaluation Study of the Training and Visit (T & V) scheme in Haryana', *Economic and Statistical Organisation*, Chandigarh.

Government of India, 1985, 'A Model Manual for Drought Management', Ministry of Agriculture, New Delhi.

Government of India, 1985, 'Statewise and Cropwise Estimates of Value of Agriculture', Central Statistical Organisation, Ministry of Planning, New Delhi.

Government of India, 1986, 'National Seminal on Training and Visit System of Agricultural Extension, Dec. 29-30, 1986—Agenda notes', Directorate of Extension, Ministry of Agriculture, New Delhi.

Government of India, (1989) 'The Drought of 1987: Response and Management', Vols. I and II, Ministry of Agriculture, New Delhi, Dec., 1989.

Parthasarthy B. *et al.* (1987) 'Drought/Floods in the summer Monsoon over different Meterological sub-divisions of India for the period 1871-1984', *Journal of Climatology*, Vol. 7, 57-70, (1987).

Sharma, R.K. (1990) "*Role of Farm Extension In Drought Management*". Agricultural Economics Research Centre, University of Delhi.

Sinha, B.P.C., (1989) 'Groundwater in Drought—Review of Indian Scene', in 'Inter-regional Symposium on Groundwater Resources, Management in Drought Prone Areas' 27 November—1 December, New Delhi.

Haryana Agricultural University, 'Package of Practices for Kharif Crops' (Issued every year).

Haryana Agricultural University, 'Package of Practices for Rabi Crops' (Issued every year).

Haryana Agricultural University, (1987) 'Choice and Protection of Crops in the Event of Drought or late Rains', July 1987, (4-page pamphlet in Hindi).

Haryana Agricultural University, (1987) 'Measures for Fighting Drought in Haryana', September 1987 (7-page pamphlet in Hindi).

Haryana Agricultural University, (1987), 'How to obtain Good Output of Rabi Crops in Drought', Nov., (1987) (8-page pamphlet in Hindi).

ANNEXURE

A copy of the pamphlet 'Choice and protection of crops in the situation of drought or late rainfall' issued by the Haryana Agricultural University during July 1987 (English translation of pamphlet issued in Hindi).

In Haryana the normal time for the onset of monsoon is between the last week of June and first week of July. However, this year monsoon rains have not started so far and there is a situation of serious drought. In the irrigated areas also there has been very little planting of paddy and the area already planted is drying due to deficiency of water. The sugar cane and cotton crops are also very weak. Fodder crops are also drying.

If monsoon arrives now, about one-and-half months of monsoon season remains. Monsoon ends by mid-September. In this situation the farmers would benefit from the adoption of following practices :

1. In rice growing areas sow short duration verities of rice or 'basmati' rice. Place 2 or 3 seedlings at one place. Paddy planted after first week of August would not mature properly.
2. In the western part of State, bajra can be sown up to the first week of August.
3. Pulse crops—Urad, Moong, Lobia, Moth and Gowar—can be sown up to the third week of August.
4. Green fodder craps—Maize, Jowar, Bajra, Lobia and Gowar can be sown during whole of August.
5. If monsoon arrives towards the end of August, sow Toria, conserve water on the fallow land and sow Raya, Sarson, Gram, Barley etc. crops during Rabi season.
6. If monsoon rains do not come at all, use irrigation properly. Note the following suggestions :

   (a) There is no need for keeping too much water in paddy fields.

   (b) On cotton fields lot of water can be saved by making ridges and channels.

   (c) Do not allow growth of weeds in fields. Soil moisture

evaporation can be considerably reduced by periodic weeding and hoeing.

7. Maximum area in Rabi should be devoted to crop requiring less water like 'Raya', Mustard, Gram, Barley etc. Besides grow less water intensive varieties of wheat.

Directorate of Research
Haryana Agricultural University
Hissar

39

Drought Disaster Management for Environmental Security and Sustainable Agriculture in India

Tapeshwar Singh

Introduction

India is rich in natural resources yet confronted with many problems including agro-meteorological one. Notwithstanding abundant natural endowments, Indian economy is often considered 'gamble on the monsoon rains', as the monsoon activity controls a major quantum about (80%) of the annual rainfall. With a large area (95 Mha) representing more than 65 per cent of net sown area still being cultivated under rainfed conditions, the impact of weather conditions during the cropping season plays a dominant role in deciding the size of cultivated net area and the productivity of many important crops, especially coarse cereals, pulses, oilseeds and rainfed rice grown in eastern India. With the increasing demand for food to feed the fast growing population, the pressure on the rainfed lands is on the increase (Singh, 2000). To enhance and sustain the increased productivity from the rainfed region including benign climate, population explosion offsets the advantages. India has achieved tremendous success in food grains output during and after green revolution, but in recent years production is stagnated at around 200.0 million tonnes per annum. This has redrawn the attention of scientists and policy makers compelling them to give a serious consideration to environmental issues such as nutrient loss, soil erosion, land conversion, soil and water degradation, loss of biodiversity, socio-cultural impacts on the environment, changes in agricultural landscape, deforestation, greenhouse gases etc. Indian region has a high degree of climatic variability. For centuries, large-scale variations in climate particularly rainfall, shaped the socio-economic conditions, cultural ethos and demographic trends of Indian subcontinent. Agriculture is the major economic activity, which is immediately and directly affected by the performance of monsoon rainfall. The well being of the farming community in the drought prone area is a matter of grave concern for the planners and policy makers in the country today. Drought is a recurring phenomena and its incidence cannot be prevented. Drought occurs in both high as well as low rainfall areas. It is a creeping phenomenon and an inevitable and unavoidable part of '*karma*'. Drought in successive year or prolonging continuously for two or more years, have disastrous effects on economy and society, and sometimes it becomes extremely fatal. Drought means a lockout in agricultural economy of the country and a serious threat to sustainability of the agro-ecosystem. Monsoon variability means socio-economic instability in India.

Precipitation Characteristics

The average annual rainfall in India is about 1200.0 mm, which is the highest rainfall per unit area in the world for a country of comparable size. The seasonal distribution of rainfall is: pre-monsoon (March-May): 13 per cent; south-west monsoon (June-September): 74 per cent; post-monsoon (October-December): 10.0 per cent; and winter (January-February): 3.0 per cent (Ramakrishna *et al.*, 2000). On the basis of average annual rainfall, the sown area in the country can be classified as follows: low rainfall region (< 750 mm) : 33 per cent; medium rainfall (750 to 1125 mm) : 35 per cent; high rainfall region (1125-2000 mm) : 24; and very high rainfall region (> 2000mm) : 8 per cent. This classification indicates that about 68 per cent of the sown area falls in the low and medium rainfall category.

Rainfed Agro-Ecosystem

Rainfed agro-ecosystem constitutes the major cropped area of the county. Out of 142 Mha net sown area, nearly two-thirds, i.e. 90 Mha is under rainfed agriculture. About 67 Mha of rainfed area falls in the mean annual rainfall range of 500-1500 mm where the productivity and stability of rainfed crop are low. There is no dependable source of irrigation for even 10 per cent of the cropped area. About 300 million people accounting for nearly 33 per cent population in these fragile areas are vulnerable to drought. It is the home of seven million households of small and marginal farmers and agricultural labourers, which is the poorest segment of our population. This agro-ecosystem is also the home of many important crops, especially coarse cereals, pulses, oilseeds and rainfed rice. With the increasing demand of food to feed the burgeoning population and stagnation in food grain output growth in the green revolution area, the pressure on these lands is already on increase. This region is economically weak, infrastructurally poor and climatically unsafe. Despite being potential growth sectors in our national economy, these areas of submarginal subsistence still suffer chronically from ecological and social disadvantages. Rainfed agro-ecosystem, which contributes to about 45 per cent of agricultural production, is entirely at the mercy of the monsoon activity over the country and is beset with problems of mid-season drought and associated impacts on crop productivity (Ramakrishna, 2000). Therefore, the proper understanding and efficient utilization of the natural resources, especially the weather, is very important for environmental security and sustainable agriculture. Weather variability also affects the appropriateness of technology and management measures in the irrigated tracks. Climate is the least manageable natural resource, yet a better understanding of its nature will be useful to develop climate/ weather based drought management strategies that will enhance benefits and create sustainability in development. Crop-weather studies will become more important tool as we embark upon the goals of increasing productivity and environmental security in the era of impending global climate change. Sustainable management of natural resources would, therefore, herald environmental stability not only in the drought prone areas but also in the rest of the country and create resilience to drought disaster.

Research and Development Infrastructure

The understanding of the role of weather and climate in influencing the crop production has considerably improved. The agrometeorological studies in India date back to 1930s with the establishment of a Division of Agricultural Meteorology in 1932 in the Indian Meteorological Department (IMD). The pioneering work of L.A. Ramdas led to the appreciation of the need for development of agrometeorological research. IMD has made significant achievement in utilizing weather information for agricultural purposes in providing crop yield forecast; conducting research on drought climatology and rainfall characteristics to determine the commencement of sowing time and agro-climatic classification for assessing the crop production potential on spatial scale. These efforts have been supplemented by the Department of Science and Technology (DST) through the establishment of a National Centre for Medium Range Weather Forecasting (NCMRWF) in 1988 with the objective of development of operational medium range (3 to 10 days) weather forecasting capability and to provide agromet advisories to the farming community in 127 agro-climatic zones of the country. Presently, the project, which came into field operation since 1991, has established about 80 Agromet Advisory Field Units, and in collaboration with All India Coordinated Research Project on Agrometeorology (AICRPAM), it is attempting to develop operational agrometeorological models for use in agromet advisory services to farmers and to assess the correctness of weather forecasts and quantify the economic impact of the a gromet advisories (Ramakrishna *et al.* 2000). In addition to these programmes, a strong research effort has been initiated under the Indian Climate Research Programme (ICRP) to analyze the climate variability and its impact on food productivity. DST is also supporting many research projects of this nature in various institutes and universities with active support from Indian Institute of Tropical Meteorology (IITM), Indian Institute of Sciences, IMD, Indian Council of Agricultural Research (ICAR), Indian Institute of Technology (IIT), Central Arid Research Institute (CAZRI), International Crop Research Institute for Semi-Arid Tropics (ICRISAT) Hyderabad, CRIDA, JARI, DWMR, PDCSR, WTCER, CARI, NVSSLUP etc. The Indian Institute of Public Administration (IIPA) has also initiated research efforts on disaster management through the National Centre of Disaster Management (NCDM). Efforts by these institutions have led to many significant contributions including the delineation of meteorological regions and agroecosystems, soil-climate zones, as well as application of water balance models quantifying the soil moisture, crop growth and yield, and research on nature of drought occurrence etc. Collaborative efforts under Indo-USAID sub-project on agro-meteorology in 1989 by providing backstop research in agro-meteorology also contributed significantly to the agro-climatic research and application of weather information for evolution of appropriate drought management techniques and strategies, optimizing cropping patterns, systems modelling etc. In addition several international organizations like WMO, FAO, UNEP, UNESCO, WCRP, GCTE, START, WCP, IHDP, IGBP, CIPS, IGU etc., have directly or indirectly made significant contribution in promoting research in agroclimatology for the benefit of the farmers and policy-makers for their resolve to sustainable agricultural development and environmental security which could mitigate drought and desertification.

Concept of Drought

The concept of drought varies from discipline to discipline. A variety of droughts are reported in literatures. Subrahamanyan (1967) identified mainly five kinds of drought such as Meteorological, Climatological, Atmospheric, Agricultural and Hydrological. Therefore, it is extremely difficult to define drought universally. Generally, drought is viewed as one of meteorological origin. It is insufficiency of soil moisture due to deficiency of rainfall that affects the crop growth adversely. In India, an area is considered to be affected by drought, if it receives total rainfall from the south-west monsoon, less than 75 per cent of the normal. It is further specified as moderate and severe if seasonal deficiency is in between 26 to 50 per cent or greater than 50 per cent of the normal respectively. The year is considered to be drought year if the area affected by any kind of drought is more than 20 per cent of the total area of the country.

Drought Prone Areas

In India, there is a long history of identifying drought prone areas. India Meteorological Department, Irrigation Commission and different committees appointed from time to time by the Government of India and the State Governments have taken into consideration certain criteria for identifying drought prone areas. These criteria are: meteorological data, *annawari* estimate and revenue remissions, the frequency of famine and scarcity and availability of irrigation facilities. However, from the analysis of annual and south-west monsoon rainfall, drought prone areas are identified. Indian Irrigation Commission (1972) has identified the drought affected and chronically drought prone areas as follows:

(i) *Drought Prone Areas :* Gujarat, Rajasthan, Punjab, Haryana, West UP, and West Madhya Pradesh, Madhya Maharashtra, Interior Karnataka, Rayalseema, South Telangana and parts of Tamil Nadu, a small portion of North-west Bihar and adjoining East Uttar Pradesh, a small portion of North-east Bihar and adjoining portion of West Bengal.

(ii) *Chronically Drought Affected* : West Rajasthan and Kutch area.

Table 39.1 : Extent and Intensity of Major Droughts of India

Years	*Rainfall Deficiency In percentage*	*Area affected (x 10 km²)*	*% Area affected*	*Total Foodgrain Output in Million Tonnes*	*% Reduction in food production from previous peak year*
1877	NA	2.03	64.7	NA	NA
1899	-26	1.99	63.4	NA	NA
1901	-16	0.89	28.5	NA	NA
1905	-15	1.09	34.7	NA	NA
1911	-14	0.97	30.8	NA	NA
1918	-23	2.16	68.7	NA	-32.2
1920	-15	1.22	38.8	NA	NA
1941	-15	1.01	32.3	NA	NA
1951	-13	1.04	33.2	NA	NA
1965	-15	1.35	42.9	72.35	-18.8
1966	-12	1.01	32.3	74.23	-15.13
1972	-25	1.39	44.4	97.03	-7.7
1974	-11	0.92	29.3	99.83	-8.59
1979	-15	1.24	39.4	109.70	-17.0
1982	-13	1.04	33.4	127.52	-8.59
1986	-13	0.60	19.0	143.42	-3.9
1987	-19	1.55	49.2	138.51	-9.0

NA = Not Available.

Source : The Table is prepared by the author.

However, drought occurrence is not confined to these areas only, though less frequently, it also affects humid and per-humid parts of the country for example, Jammu and Kashmir, Assam, Manipur, Orissa etc. Irrigation Commission has identified 72 districts either partly or wholly in the country as drought prone. However, at present, these project areas of Drought Prone Area Programme (DPAP) are spread over a large part of the country comprising 615 blocks in 91 Districts of 13 States.

Probability of Drought Occurrence

Most of the drought prone areas fall either in arid or semi-arid regions. The probabilities of occurrence are high in the arid regions of west Rajasthan compared with other meteorological sub-divisions. The extent to which the monsoon rainfall occurred over the country during the drought years compared to average annual rainfall for the years 1880 to 1986 showed that lowest amount of rainfall occurred during the year 1899, 1918 and 1972. Droughts occur at random and no periodicity has been noticed (Table 39.1). The probability of occurrence of drought in the country, as reported by Upadhyay (1994) has been shown in Table 39.2.

Table 39.2 : Probability of Occurrence of Drought in Different Meteorological Sub-Divisions

Meteorological Sub-division	*Frequency of Deficient Rainfall (75% of normal or less)*
Assam	Very rare, once in 15 years
West Bengal, Madhya Pradesh, Konkan, Bihar & Orissa	Once in 5 years
South Interior Karnataka, Eastern Uttar Pradesh & Vidarbha	Once in 4 years
Gujarat, East Rajasthan, Western Uttar Pradesh	Once in 3 years
Tamil Nadu, Jammu & Kashmir & Telengana and West Rajasthan	Once in 2.5 years

Source : Apparao (1991).

Drought Impact

Drought is the main precursor of famine. The impact of droughts on crop productivity varies from crop to crop and place to place according to the agro-meteorological and socio-economic conditions. India is blessed with a benign climate and receives substantial amount of rainwater through monsoon rain, eventhough, some parts of the country experience frequent drought. Historical evidences suggest that our country had been frequently affected by drought, which created famine in large parts of the country on several occasions. Drought takes a heavy toll of life and property and offsets environmental stability, agricultural sustainability and food security. During nineteenth and twentieth centuries, drought caused several scarcities and famines resulting in heavy starvation deaths, social distress and widespread damage to national economy. For example, 1 million people died in famines between 1800-1825 AD, 0.4 million died between 1825-1850, 5 million died between 1850-75 and 15 million died between 1875-1900 (Das, 1981). The 'Orissa Famine' of 1865 alone starved 1,35,000 persons to death. The bulk of livestock die during the droughts. The extent and intensity of major droughts is shown in Table 39.1. Therefore, any adverse environmental factor that affects the agriculture, consequently influences the national economy and food security, because nearly 33 per cent of the gross national income is still derived from agriculture and 70 per cent of the country's population still depends on it for the livelihood. The impact of drought can be broadly categorized into two sections:

(i) *Environmental (Natural) impact :* This includes reduction in surface (Stream flow) and sub-surface water reservoirs, destruction of various ecosystems and reduction in natural vegetation cover; and

(ii) *Cultural impact :* Which includes economic (agricultural crops, livestock, fishing, industry etc.), social conflict, outmigration of livestock and human population, starvation, undernourishment, malnutrition, unemployment, political-change in state regimes etc. Drought impacts and the socio-economic damages can also be seen in the sections on drought management policy and drought in historical perspectives.

Records of drought and famines are also available from the chronicles of Mughal India. Historian Quadir Badaoni mentioned the drought of

1555-56, which severely affected parts of northern India particularly the neighbourhood of Delhi. It created such a bad situation that cannibalism was witnessed. The failure of rain in 1630-31 resulted in famine in Gujarat where people began to devour each other and flesh of a son was preferred to his love (Singh, 1978). Drought calamity of 1899 resulted in the worst famine on record in which the annual rainfall for the country as a whole was 26 per cent below normal. Sir John Elliot the then government meteorologist afterwards estimated that this drought was the greatest in extent and intensity, which India has experienced during the last 200 years. Thus, drought has been a repetitive calamity and is unavoidable.

Historical Development of Drought Management Policy

Drought relief has been in vogue in India from the very beginning of the human settlement. For instance, Kautilya's *Arthashastra* written more than 2000 years ago pronounced famine relief of various forms in the past. The chief methods of famine relief adopted by Indian rulers included free distribution of raw food grains, opening of free kitchens, opening of public grain stores to the people, remission of the taxes, construction of public works, canal and embankments, digging of wells, encouragement of migration and increase in pay of soldiers (Pant, 1992). However, there is no substantial historical statistical evidence to judge the real efficacy of relief efforts prior to the 19th century. The principle of doles of food and money to the affected population was a common practice. However, there has been a noticeable change in drought management approroach in post independence India. The government no longer remains either a passive witness or an unmoved spectator to the miseries of the affected population. As reflected from the drought management of 1987 as soon as drought manifests, the governments gear up the entire state machinery to meet the challenge at war footing and provide prompt relief to the sufferers. For an appreciation of the famine management policy it is worthwhile to recount some of the important stages in its evolution (GOI, 1989).

The famine policy of the earlier rulers in India evolved through a process conditioned by the resources of the State, the philanthropy and munificence of the rich and the political and economic philosophies of the times. The relief policy of these times was, broadly speaking *adhoc* and not based on any definite principles. This was due to the stage of political development, lack of resources and communications, and the inadequate appreciation of the vulnerability and the needs of the affected population. In the era preceding the British rule, the village communities were mostly self-sufficient and the means of communications were virtually non-existent. Consequently, trade and movement of food grains even in times of plenty were limited. The emperors are said to have maintained stores of food grains in their capital as war chests, and richer landlords also kept surplus from the years of bounty as insurance against years of scarcities. Depending upon the benevolence of the rulers and the severity of the famines, grains were either sold or distributed free. The Gujarat relief manual recounts at length the history of Indian famines as noted by A. Loveday in his book—*History and Economics of Indian Famines* (1914) and has on record that Mohammed-bin-Tughlaq was the first ruler to take vigorous measures to alleviate the effect of the drought that occurred in 1343 AD during his regime. In addition to the distribution of 6 months supply of food grains to the inhabitants of Delhi, advances were made from the treasury not only for the cultivation of land but also for digging of wells. In AD 1630, Shah Jehan is said to have distributed Rs. 5,000 every Monday to the deserving poor, and Rs. 50,000 in Ahmedabad during the famine. The same principles of doles of food was also adopted and land revenue was subsequently remitted to the extent of Rs. 30 lakh. A departure was, however, made in that the relief was not confined to the capital of Delhi, but it was extended to Burhanpur, Ahmedabad and the country around Surat, where some kitchens and alms-houses were established. In AD 1596 some relief works were established and the strength of the army was increased to support the poor.

According to Loveday, the important famines between AD 297 and 1907 tended to recur in cycles of five years and the greater ones in cycles of 50 years. Roughly speaking, it was towards the middle and the end of each century that the most disastrous calamities occurred in India. The famines of AD 1343-45, 1540, 1630, 1745 and 1837 were all intense in certain districts though comparatively limited in area and short in duration. Those of AD 1396, 1596, 1660-61, 1803-04, 1896 and 1900 were the most disastrous and widespread famines.

In spite of the fact that such a large number of famines occurred in India over the past centuries, the relief policy of Government was torn between the conflicting goals of saving life and of securing the maximum possible economy in relief expenditure. Even during the terrible famine of 1877, the British government warned the special officer Sir Richard Temple that "the task of saving life irrespective of the cost is one which is beyond their power to undertake". However, some officials of nineteenth century took keen interest in solving the problem. Colonel Baird Smith recommended the extension of the irrigation system, improvement of communication, and remission of land tax in 1860-61. The Famine Inquiry Commission under General R. Stratchey (1878) said "The paramount duty of the State was to offer protection to the people of India from the effects of uncertainty of the seasons".

Famine Commissions

Three successive Famine Commissions were appointed during the period from 1878 to 1900 to investigate different aspects of famines such as nature and causes of recurring famines, assessments of relief measures being taken and guidelines for the future. The employment of the affected people on large works of permanent utility, improvement in internal communications, extension of irrigation facilities, establishment of Famine Insurance Fund, framing of Famine Codes for Provincial Governments, suspension and remission of land revenue, well defined programmes of public and village works, liberal advances to the farmers in the earlier stage of distress, welfare of deserted children, preservation of cattle, cultivation of fodder crops, opening of cattle camps and involving NGOs for administering relief were some of the important recommendations, which later on formed the basis of drought mitigation schemes.

Famine to Drought

The historical evolution of drought management policy witnessed a noticeable change in approaches of the drought management policy to the famines of the pre-independence days and to droughts, which were experienced since then. Schematic presentation of major components of drought management in India is shown in Table 39.3.

Table 39.3: Major Elements of Drcught Management in India: A Historical Perspective.

Approach	*Famine Relief*	*Scarcity Relief*	*Drought Relief*	*Drought Management*
TIME PERIOD	Pre-Independence Era	Post-Independence and up to mid-1960s	From late 1960s to mid-1970s	Late 1970s to present
OBJECTIVES	Minimise starvation deaths	Prevent starvation deaths	Ensure economic access to food	Preserve quality of life
TRIGGERS	-	Social distress indicators	Economic distress indicators	Agro-hydro stress indicators
INTERVENTION	Very late	End of harvest season	End of monsoon season	Within the monsoon season
PREPAREDNESS	Absent arrangement	Administrative network	Food distribution, drinking water plans	Contingency crop/fodder
KEY MANAGEMENT PROGRAMMES	Limited intervention to minimise deaths	Free kitchens, gratuitous relief	Employment generation programme	Crop stabilization
ADMINISTRATIVE APPROACH	Famine code	Scarcity relief manuals	Drought relief manuals	Drought handbooks
EPISODES	Pre-Independence Famine management	1965 to 1967 drought	1972 drought	1979 and 1987 droughts

Source : A.R Subbiah (1993).

The Famine Relief Codes of the erstwhile provinces were replaced in the fifties by the Scarcity Relief Manuals of various States which describe 'scarcity' as a marked deterioration of the agricultural season due to the failure of rains or floods or damage to crops from insects resulting in severe unemployment and consequent distress among agricultural labour and small cultivator. In assessing the state of a season for the purpose of determining whether scarcity is to be declared or not, factors such as *annawari* of crops in the current and in the previous two years, the availability and prices of food grains and fodder, the state of employment and trends in wages, unusual movements of labour from rural areas, the state of crime and other factors indicating signs of distress such as malnutrition among children, were taken into consideration.

Scarcity manuals lay down measures to provide reasonable purchasing power to the affected people as well as to ensure availability of goods and services. The more recent drought manuals also lay down measures not only towards alleviating distress of the affected population but also to bring out alternative arrangements in maintaining crop production through contingency planning and better water management and development efforts by creating permanent assets through employment generation works to the extent necessary.

New Drought Codes

Every major drought contributed in bringing about qualitative improvements in drought management policy. Drought of 1965-66 contributed to building up of a reliable public distribution system (PDS) to take care of the food emergencies. The privations suffered during this drought also spurred the country into embarking upon certain fundamental changes in the agricultural strategy, which ultimately ushered in a Green Revolution making the country self-sufficient in foodgrains production. The drought of 1972 focused on the need of evolving massive employment generation programmes for enhancing the purchasing power of the people rather than running free kitchens, while the drought of 1979 underlined the need of creating durable and productive assets for enabling the people of the affected area to withstand onslaught of future droughts with greater resilience. A noticeable shift in thinking occurred from curative measures to preventive against adverse effects of drought.

The drought of 1987 focused not only on the needs of providing access to food to the affected people but also maintaining the quality of life. It is no more a question of saving the affected people from the threat of widespread starvation during such a natural calamity. During 1987, drought management was reoriented towards employment generation, provision of drinking water supply, adequate health care, creation of food security, fodder availability and supply of essential commodities besides various drought-proofing schemes. It also emphasized on the prevention of environmental degradation and creation of rural infrastructures. Despite many efforts inadequacies in drought management policies were felt. The Government of India (1994) appointed a committee under the Chairmanship of Hanumantha Rao. The report recommended for implementing these programmes on watershed basis. The main thrusts of the DPAP were in the direction of proper ecological balance, ecologically integrated development, development and rational management of irrigation resources, soil and water conservation and afforestation; restructuring of cropping pattern and pasture development, changes in agronomic practices, livestock development, provision of drinking water supply, development of rural communication, and development of small/ marginal farmers and agricultural labourers (Singh, 1995). The central theme of relief today is to meet the food and nutritional needs of all sections of the people keeping in view their normal energy requirements, supply of drinking water, providing adequate heath care and fodder for the cattle.

The drought management policy presently under implementation seeks to provide for the social and economic goals of the welfare state and the egalitarian objectives of the government as embedded in the Constitution of India. The objective is not only that no one should die of starvation but more importantly to prevent physical deterioration and destitution of the people and to enable them to resume normal pursuits to encourage life at the earliest. It also aims to encourage the community efforts by sharing a common concern and to fashion or shape the events and social infrastructure in such a manner that the recurrence of scarcities and other impacts are minimized. The approach of the present drought management policy, therefore, is more towards drought mitigation and is preventive in nature rather than merely curative. The relief measures are, therefore, not to be conceived in isolation but integrated with the development ethos and programmes under implementation under various Five Year Plans in India.

Drought management strategies in India are comprehensive and to some extent they have succeeded in creating strong resilience to drought impact Table 39.2 and also in improving knowledge of preparedness, mitigation, prevention, response, recovery and development. However, adequate attention has not yet been paid on the agroecological and environmental considerations. Therefore, the proper understanding and efficient utilization of natural resources, especially (rainfall) though it is meagre and soil and vegetation etc. is of great concern for the improvement and sustainability of agriculture of the rainfed/drought-prone agro-ecosystem.

Table 39. 4 : Comparison of Impact and Distress Indicators between 1965 and 1987

	1965	*1987*
A. Impact Indicators		
1. Geographical area affected (%)	54.00	63.00
2. Total population affected (Million)	40.76	282.41
3. Population affected (%)	19.00	21.00
4. Rainfall anomaly (%)	-18.60	-19.00
5. Crop area affected by drought (% of sown area)	53.23	53.66
6. Fall in foodgrain production over previous peak years (%)	17.01	12.02
7. Number of meteorological zones affected	19	21
8. Number of drought years preceding	0	3
B. Distress Indicators		
1. Net availability of foodgrains per capita per day (in grams)	401.40	447.60
2. Import of foodgrains (in million tonnes)	10.60	0.00
3. Average calorie intake per person per day	1400.00	2000.00
4. Starvation level (% of households whose calorie consumption fell below 500 kelocalories)	37.00	0.20
5. Rise in foodgrain price (%)	65.00	10.00
6. Increase in petty crime (%)	39.20	Negligible
7. Sale and mortgage of land etc. (% household)	5.53	Negligible

Source : Compiled by the author from various reports.

Main Ingredients of Drought Management

At the recommendation of the Task Force set up by the Planning Commission, Government of India, in October 1971, submitted its report in 1973 containing detailed tasks and components of development strategies, which are given hereunder:

(a) Restoration of ecological balance,
(b) Development and management of water resources,
(c) Soil and water conservations and afforestation,
(d) Restructuring of cropping pattern and pasture.
(e) Changes in agronomic practices,
(f) Livestock development,
(g) Provision of drinking water supply,
(h) Development of rural communication, and
(i) Development of small and marginal farmers and agricultural labourers.

In view of the recommendations, the design and the content of the DPAP and Desert Development Programme (DDP) emphasized on: agro-meteorological studies, pasture development, agro-forestry, irrigation and soil conservation and drought proofing techniques evolved and used by the farming community. Dry Farming techniques based on traditional knowledge used by farmers such as agronomic practices, supplementary/life saving irrigation, farm ponds, intercropping, proportionate cropping, alley cropping, crop rotation, mulching, watershed management etc. as well as other measures like water budgeting, Bombay Dry Farming System, conservation agronomy, early warning, contingency crop planning etc. have also contributed significantly to the drought management in India. Traditional strategies such as moisture security, biomass stability, collective sustenance, flexibility and diversifications are the dominant strategies adopted by farmers through which they historically sustained themselves in climatically unstable dry tropical areas. Hanumatha Rao (1994) after reviewing the ongoing programme contents of DPAP and DDP recommended the reinforcement of research and training programmes for policy implementers and the beneficiaries, conservation of natural resources, use of traditional knowledge of farmers, involvement of NGO, creation of adequate infrastructures, watershed management etc. This committee duly emphasized the management approach for drought proofing. Rainwater management is a prerequisite for sustainability in agricultural development as well as sustainability of the ecosystem as a whole. Fortunately, India is blessed with very rich potential of farmer's traditional knowledge including rainwater harvesting and area specific dryfarming techniques. The model manuals of drought management have incorporated broadly: (i) early warning, (ii) contingency planning, (iii) employment generation, (iv) covering vulnerable section of the population, (v) strengthening the public distribution system, (vi) water supply, (vii) medical treatment, and (viii) relief measures for livestock.

Notwithstanding the benefit of planned socio-economic development of the country as a whole under various Five Year Plans, drought management has created a substantial resilience in the impact of drought not only at household level but also at national level. The comparison of impact and distress indicators between 1965 and 1987 clearly indicates that drought management measures have been very effective and successful in mitigating drought and creating strong resilience to drought impacts (Table 39.4).

However, there should not be any complacency in drought management policy implementation as well as rejuvenation, updating and reorientation. The need of the time is a comprehensive National Drought Policy but still it is a far cry. However, the National Water Policy (GOI, 1987) had given some place to the drought management in section 18 which said that "Drought Prone Area should be made less vulnerable to drought associated problems through soil-moisture conservation measures, water harvesting practices, the minimization of evaporation losses, the development of the ground water potential and the transfer of surface water from surplus areas where it is feasible and appropriate. Pastures, forestry or other modes of development which are relatively less water demanding should be encouraged. In planning water resource development projects, the needs of drought-prone areas should be given priority. Relief works undertaken for providing employment to drought stricken populations should preferably be for drought proofing." The Indian Council of Agricultural Research (ICAR), through its constituent Institutes and the All-India Coordinated Research Projects in collaboration with the State Agricultural University developed useful technological inputs which are responsive to varied cropping scenarios and production systems and put the required dynamism into agricultural production strategies, thus enabling the country to tide over the difficulties created by drought situations.

National Agriculture Policy 2000

After a long debate since 1992 National Agriculture Policy has finally been framed by the Department of Agriculture, Government of India and announced in July, 2002 by the present government. This document under sections 41 and 42 has, dealt with risk management, which is referred below:

> Despite technological and economic advancements, the condition of farmers continues to be unstable due to natural calamities and price fluctuations. National Agriculture Insurance Scheme covering all farmers and all crops throughout the country with built-in provisions for insulating farmers from financial distress caused by natural disasters and making agriculture financially viable will be made more farmers specific and effective. Endeavour will be made to provide a package insurance policy for the farmers, right from sowing of the crops to post-harvest operations, including market fluctuations in the prices of agricultural produce.
>
> In order to reduce the risk in agriculture and impart greater resilience to Indian agriculture against droughts and floods, efforts will be made for achieving greater flood proofing of flood prone agriculture and drought proofing of rainfed agriculture for protecting the farmers from vagaries of nature. For this purpose, contingency agriculture planning, development of drought and flood resistant crop verities, watershed development programmes, drought prone areas and desert development programmes and rural infrastructure development programmes will receive particular attention.

Thus, apart from continuation of various drought proofing strategies and techniques already going on in the area, synergism among ecological balance, environmental security and sustainability had to be on the priority agenda of the development process in the ensuing years of the 21st century. Therefore, the current drought mitigation measures adopted by NCDM (2000) on the occasion of the National Day of Disaster Reduction has given priority to the following programmes :

1. Arrangement for reasonable buffer stock of food grain and fodder.
2. Ensure supply of potable water in rural areas for human and livestock in drought-affected areas.
3. Assess fodder requirements in drought affected districts and locate areas where the shortage is likely to occur and arrange supplies from outside.
4. Fodder cultivation to be encouraged wherever feasible.
5. Rejuvenation of traditional rainwater systems, viz., *Nadis, Tankas, Khadins,* etc.
6. Rainwater harvesting for both the drinking and cropping.
7. Management of human livestock population to reduce pressure on fragile arid ecosystem.
8. Timely availability of credit, postponement of Revenue collection and repayment of short-term agriculture loans.
9. Appropriate land-use planning (inter-cropping system), discouraging water intensive crops, encouraging sprinkler and drip irrigation systems.
10. Creation of Local Task Force in each district to initiate relief measures immediately after the drought takes place.
11. Implementation of crop and livestock insurance schemes.
12. Provisions for cattle camp in drought affected areas.
13. Early warning and drought monitoring on the basis of long, medium and short-term forecast.

Conclusion

India has to some extent succeeded in eliminating food scarcity and hence famine in different parts of the country. The impact of drought has been reduced and drought management strategies have significantly increased the country's resilience to drought disaster but regional and household or farm-level vulnerability still persists. Considerable progress has already been made but much has still to be done. We need to emphasize on certain strategies such as understanding of the weather pattern and the vulnerability of the region, efficient management of rainwater, watershed management, development of response farming strategy, NGOs and the community participation and awareness, strengthening communication system, environmental management, drought monitoring and harnessing the tremendous potential of the traditional knowledge of farming community under the drought management strategies, which can protect such regions from being frequently subjugated to the onslaught of drought. Different development programmes must not be taken in piece-meal manner, rather, on permanent basis for holistic development of the region while preserving the environmental potential for sustainable agricultural development.

REFERENCES

Alexander, David. (1993) *Natural Disaster*, UCL, pp. 144-54.

Apparao, G. (1991) Droughts and Southwest Monsoon. Training course in monsoon meteorology, *3rd WMO, Asian/African Monsoon Workshop*, Pune India.

Bhamle, H.N. and Mooley, D.A. (1997) Large scale droughts/ floods and monsoon circulation. *Monthly Weather Rev*, Vol. 108, pp. 1197-1211.

Blanford, H.F. (1986) Rainfall of India, *Mem. LMD.*, 3, p. 658.

Carter, W.N. (1996) Disaster management: A disaster manager's handbook. *Asian Developmet Bank*, Manila, p. XXIII.

Chaudhary, T.N. (1995) India (1) Report of Productivity Council, Tokyo: *(Asia Productivity Council)* Organisations Seminar, July 1216, 1995, Colombo on 'Agricultural Water Management Technology in Asia and Pacific' Pub. By A.P.O. Tokyo, Japan, pp. 15869.

Das, P.K. (1981) A historical perspective: Drought and famine in India. In: Pearce, R. (ed.) *TropicalDrought: World Meteorological Organization (W.M.O.)*, Geneva, pp. 3-12.

Devindra Sharma. (1997) *Times of India*. New Delhi, June 27-28.

Govt. of India. (1973) *The Task Force Reports on Integrated Rural Development*, appointed by the Planning Commission under the chairmanship of Dr. B.S. Minhas, New Delhi.

Govt. of India. (1989) *The Drought of 1987: Response and Management*, Vols. I and II, Department of Agriculture and Cooperation. Min. of Agriculture, New Delhi.

Govt. of India. (1994) *Reports of the Technical Committee on Drought Prone Areas Programme and Desert Development Programme*. Min. of Rural Development, New Delhi.

Govt. of India. (1993) *Guidelines for Watershed Development*, Min. of Rural Development, New Delhi.

Govt. of Rajasthan. (1993) Perspective Plan for Drought Management, Jaipur.

Haq, A.K. and Samarsekera, B.M.S. (1995) Current Development in Applying Water Management Technologies, *Report of Asia Productivity Council*, Tokyo, Japan, pp. 43-50.

Herath, S., Katayama, T. (1993) Towards Natural Disaster Reduction. *Proc. of Workshop* of *the VII Pacific Science Inter-Congress*, June 27-July 3, 1993, Okinawa, Japan.

I.I.P.A. (1993) *Proc. India Disaster Management Training Country Workshop*, July 12-16. 1993. pp. 79-89.

Kulshrestha, S.M. (1997) Drought Management in India and Potential Contribution of Climate Prediction, *Joint COLA/CARE Technical Report No. 1*, September 1997.

Pant, J.C. (1992) Disaster Relief in India. In: Mishra, G.K. (ed.) Natural Disaster Reduction. I.I.P.A., New Delhi, pp. 59-60.

Ramakrishna, Y.S.; Rao, G. G.S.N.; Keshav Rao, A.V.R. and P. Vijoy Kumar. (2000) Weather Resource Management. In: Yadav, J.S.P. and G.B. Singh (eds.) Natural Resource Management for Agricultural Devolopment in India. *Proc. International Conference of Managing Resources for Sustainable Agricultural Production in the 21st century*, Feb. 14-18, New Delhi, India.

Schell, B. (1995) What is Disaster Relief ? Agricultural and Rural Development, Vol. 2, 1995, pp. 8-10.

Sharma, V.K. (ed.) (1995) *Disaster Management*. Indian Institute of Public Administration. New Delhi.

Shukla, J. and Paolino, D.A. (1983) The Southern Oscillation and Long-range Forcasting of the Summer Monsoonal Rainfall over India. *Monthly Weather Review, 111:* 1830-1837.

Singh, Tapeshwar. (1978). *Drought Prone Areas in India*. People's Publishing House (P) Ltd, New Delhi.

Singh, Tapeshwar. (1995) *Drought Disaster and Agricultural Development in India*, People's Publishing House (P) Ltd., New Delhi.

Singh, Tapeshwar. 1(998) Drought Disaster Management: An Indian Perspective. *National Geographical Journal of India,* Vol. 44 (14), Mar.-Dec., 1998, pp.206-15.

Singh, Tapeshwar. (2000) Drought a challenging natural hazard in sub-humid climatic region: A case study of Palamau. In: Singh, R.B. (ed.) *Disaster Management,* Rawat Publications, New Delhi. pp. 118-31.

Singh, Tapeshwar. (2000) Climate variability and its impact on crop productivity : A case study of Bihar Plateau, India. *Proc. of International Symposium on Climate Variability, and Their Impacts,* Commission on Climatology, 29th IGC, August 9-13, 2000, Seoul, Korea. pp. 124-28.

Sinha, S.K.; Kailasanathan, K. and Vasistha, AK. (1987) Drought Management in India: Steps towards eliminating famines. In: Wilhite, Donald, A.; Easterling, W.E. and Wood, D.A. (eds.) *Planning for Drought: Toward a Reduction of Societal Vulnerability*, Westview Press, Boulder, Colorado, pp. 453-70.

Siromony, P., Michal, Vetha. (2000) *Source Book on District Disaster Management*. Published by LBSNAA, Mussoorie and Ministry of Agriculture, Govt. of India, Aug. 2000.

Subbiah, A.R. (1994) Drought Management through Anticipating Multidimensional Approaches: A Case Study. In: Sharma, V.K. (ed.) *Disaster Management*, IIPA, New Delhi, p. 81.

Subrahamanyam, V.P. (1967) Incidence and Increase of Continental Drought. *WMO/IHD Report No. 2, WMO,* Geneva.

Upadhyaya, D.S. (1994) Hydrometeorology in Agriculture. *Indian Farming, Vol. 44,* No. 7. PP. 41-45.

Virmani, S.M. 1994. Climatic Resource Characterization in Stressed Tropical Environment: Constraints and Opportunities for Sustainable Agriculture. In: Virmani, S.M.; Katyal, J.C.; Eswaran, H. and Abrol, I.P. (eds.) *Stressed Ecosystem and Sustainable Agriculture*, Oxford & IBH, New Delhi, 1994, pp. 149-60.

SECTION-IV

GEO-SPATIAL TECHNOLOGY APPLICATION

40

Land Use/Land Cover Studies Using Romote Sensing Data

An Indian Experience

N.C. Gautam

Introduction

The total geographical area of India is about 329 million hectares and the total population is more than 1000 millions according to the 2001 census. India is, therefore, the seventh largest and the second most populous country in the world with amazing physical and cultural diversity.

Diversity in the physical landscape is the cumulative effort of both geological and geomorphological processes since millions of years. This has finally resulted in the formation of variety of land with its associates land systems, forms and surfacial material such as mountains, hills, valleys, plateaus, plains etc. the different types of land are subject to different types of utilization to the possible extent. Land is the most important natural resource on which all men's activities are based. Growing population pressure and human activities are increasing the demand on the limited land and soil resources both for agricultural and other land uses. This is emphasized by the decrease in the per capita availability of arable land from 0.34 ha in 1951 to 0.21 ha in 1981 and to 0.17 ha in 2001 AD. India being predominantly an agricultural country has 142 million hectares (43.4 per cent) of the total geographical area put under regular cultivation. Out of the remaining 186 million ha, about 67 millions ha is under forest cover (NLU & CB, 1988). Even here, 85 millions ha. of agricultural land

and 37 million ha of forest land are reported to be suffering from varying degrees of soil degradation. Of the remaining 56 million ha, 3 million ha are classified as tree crops or orchards and the remaining are classified into pastures (13 mha) culturable waste (17 mha) and fallow (23mha). All these lands are so over-grazed or over used that their productivity is very low. This situation can be improved in two ways :-

1. To increase the yield per ha by increasing the inputs of fertilizers, irrigation facilities, high yielding variety of seeds, through mechanization and better management.
2. To identify and bring more land area under cultivation. afforestation and plantation. Although with the event of 'Green Revolution' (since late I960s) and better agricultural infrastructure and mechanisation. India has been able to increase, not only the total food production (150 million tonnes: 1984-85) to 198 million tonnes: 2001 but also the yield per ha. But the shift from the traditional subsistence farming to increase agricultural practice has given rise to problems like increasing salinity, erosion, waterlogging etc. So the alternative is to find suitable measures to reclaim such land to make it productive again.

The Need

Planet Earth, with its basic elements such as Air, Water and Land is the only habitable place in the universe. Its natural bountiness has sustained a variety of life for millions of years. However, the stress on the balance of nature has never been so profound as it is today.

The ever-increasing population, unplanned growth, pollution and indiscriminate destruction of forests have led to land degradation, poor crop yield, drinking water problem, health hazards and pressure on urban facilities. These problems prompt continuous caution and monitoring in order to evolve developmental plans for effective management of land and water resources.

Frequent natural calamities like floods, cyclones, earthquakes, landslides etc. leave behind a trail of destruction in their wake setting all the development activities back by several paces. An effective disaster warning system is imperative to ensure minimum damage to life and property and plan remedial measures.

Since most of these problems operate in a large space and time, they require observations from a position that gives an overview of the situation.

An 'eye in the sky' would be the most ideal platform. Manmade satellites provide a reliable platform for this purpose. We are now better equipped to manage our environment wisely and effciently with the help of this ability of viewing Earth from sky—called Remote Sensing.

Remote Sensing ensures,

- Large area coverage enabling local, regional, national and global surveys;
- Continuous observation for disaster mapping and assessment;
- Periodic monitoring of environment and resources like land, water, agriculture, etc.;
- All weather viewing capability;
- Mapping inaccessible areas like flood-affected areas, mountains, swampy areas, thick forests, etc.;
- Fully computer-compatible data, ensuring fast and accurate results;
- Cost-effective assessment of environment and habitation.

Man has been interested in the deep dark umbrella called Space since time immemorial and has been trying to unravel the mysteries of the celestial bodies. The sun and moon have been used since ages to determine time and distances. As man became more and more civilized, curiosity for the mysteries of space gave way to several novel inventions in order to obtain more information.

What is Remote Sensing?

The concept underlying the remote sensing technology can be understood with the following simple analogy. When we look at the world through our eyes, we don't actually come into contact with it. Our eyes record the shape, size and colour of an object and pass on the information to the brain. Based on the knowledge already available with it, the brain then recognizes the object. Here what we are actually doing is 'sensing' think 'remotely' from a distance. Our eyes are 'sensors' that record information to pass it on to the brain. Similar is the technology of remote sensing—*obtaining information about an object without actually coming into contact with it.*

So we can think in this way that our body is a 'satellite' which moves and collects information through our eyes which are the 'sensors'. We cannot see any object in the dark. Just as our eyes need objects to be illuminated by light so that we can see them, sensors also need a source of 'light' to illuminate the Earth's surface. Different forms of light or electromagnetic energy are utilized for this purpose. Whenever light felts on an object, part of it is absorbed, part of it is allowed to pass through and the remaining is either reflected or scattered. The light that is reflected back to us gives us information about the object.

Similarly, a sensor onboard a satellite collects information that is reflected from different objects an the Earth's surface. This information conveyed, back to earth, so that we can get the required information about our abject of interest.

Where do we Place a Sensor in Space?

To carry out Earth observations, the camera should be place in space. A Platform is a stage to mount the camera of sensor to obtain information about a target. These platforms could be airborne or space-borne.

Remote Sensing Application

The sensors of IRS satellites have grown from strength to strength in respect of resolution and area coverage. In a country like India, which has diverse spreads in terms of geography and population, it is essential to put satellite data to effective use and ensure that the advantages percolate to the grassroots levels for overall development.

Management of natural resources developmental planning, environmental monitoring and disaster management are some of the key areas of concern, besides a host of applications that are being carried out in order to optimally utilize the potential of satellite data.

Specific application projects are undertaken based on user requirement.

Agriculture and Soils *(India is a agriculture-intensive country. Timely and reliable information on agriculture and soils are required for optimal land use planning and formulating policies on food security) :*

- Crop Acreage and Production Estimation (horticultural, multiple, mono)
- Mapping of soil resources, land capability and irrigability
- Soil moisture estimation
- Mapping and monitoring water-logged areas, salt-affected soils, eroded lands, shifting cultivation

Water Resources *(Two-third of Earth covered with water, yet demand for water grows rapidly; effective water resources management is the need of the hour) :*

- National Agricultural Drought Assessment and Monitoring System
- Flood/cyclone damage assessment and flood risk zone mapping
- Monitoring vulnerability of embankments of canal/tank
- Irrigation system
- Reservoir sedimentation and drainage congestion studies
- Snow cover mapping and snowmelt runoff forecast

Forestry, Ecology *(Large forest areas cleared for want of space for habitation and agriculture, irrevocable damage to fragile ecology. disappearing rare species of wildlife. Imperative to nature our forests) :*

- *Biodiversity* conservation studies
- Forest cover mapping and monitoring
- Forest timber volume estimation and stock assessment
- Wildlife habitat analysis
- Shifting cultivation areas assessment
- Forest working plan
- Grassland mapping

Geosciences *(Understanding processes of Earth that cause long-term implications. Valuable satellite data helps identify geological features, processes and anomalies)* :

- Identifying sustainable drinking water sources
- Geological/geomorphological mapping
- Mineral exploration studies
- Ground water potential zone mapping
- Geoenvironmental studies
- Mini hydel project site selection
- Reservoir site evaluation

Land use/cover *(Multiplying population, limited lands, wastelands, unplanned growth, pressure on urban and rural systems, Dynamic monitoring required)* :

- Land use/cover mapping
- Change analysis studies
- Mapping of wastelands of India
- Urban sprawl studies
- Corridor survey
- Rural roads inventory
- Inputs for master plan development of cities

Environmental/disaster Studies *(Frequent disasters in India, both natural and manmade. Quick monitoring and damage assessment imperative for suitable remedial measures, preserving environment)* :

- Landslide hazard zonation mapping
- Earthquake studies
- Environmental impact assessment studies—mining areas, oil pipeline route planning
- Forest fire damage assessment and risk mapping
- Monitoring underground coal fire
- Mapping volcanic eruptions
- Land-air-ocean interaction studies
- Atmospheric constituents and aerosols studies

Land Use Studies in India

- Pre-1984 technology demonstration studies in few states/ regions
- Project based classification
- National Level Operational Studies (From 1984)
- Land Use/Land Cover Mapping for agroclimatic zone plans.
- Leading to Land Use Census (NR Census)
- National Wastelands Inventory Project
- Leading to Monitoring of Wasteland programmes of MRD

Present Scenario of Land Use/Cover Information Generation and Usage

Presently land use information is available from the following sources:

1. Revenue records compiled by the Directorate/Bureau of Economic and Statistics (DES/BES) which is mainly tabular. Further, the 9-fold classification does not encapsulate all the land use categories as is required.
2. Topographical maps from Survey of India represent very broad land cover categories.
   These maps mainly provide topographical information.
3. Land use Atlas from NATMO are mainly small state and are secondary compilations.
4. Soil Survey Organization (NBSS&LUP and AIS&LUS) generate soil and land capability maps for specific project/ areas.

Nation-wide Land Use/Cover Mapping

No. of Districts covered	:	442 (entire county)
Scale of mapping	:	1:250,000
Database	:	IRS - LISSI (FCC's & CCT"s)
Seasons	:	Kharif (Sept. to Nov.) Rabi (Feb. to Mar.)
Methodology	:	Visual (2T4 dist.)
Year of data	:	1988-89
Work centres	:	26
Classification	:	22-fold classification

Land use/Land cover Mapping Provides

1. The study provided for the fist time the spatial pattern of land use/cover information.
2. The Net sow area is about 146 million hectares as against 142 million hectares reported.
3. The extent of grazing/grasslands is very sparse.
4. The extent of shifting as well as abandoned shifting cultivation is about 2.82 million hectares.
5. The forest type and density was mapped.
6. All the surface waterbodies and its water spread was mapped.
7. The spatial pattern and the sprawl of settlements could be analyzed.
8. The database is extensively used by Pollution Control Boards for Industrial zoning.

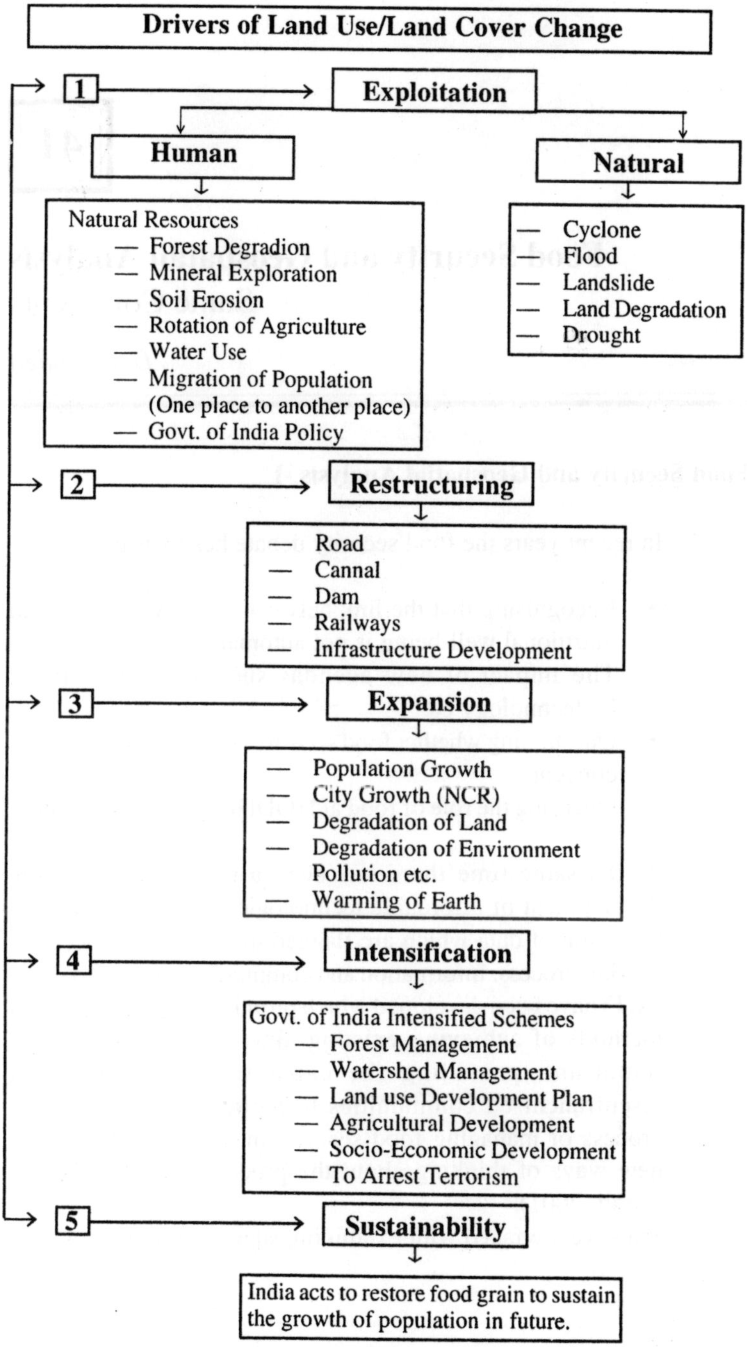
Drivers of Land Use/Land Cover Change
1
Exploitation
Human
Natural
Natural Resources
— Forest Degradion
— Mineral Exploration
— Soil Erosion
— Rotation of Agriculture
— Water Use
— Migration of Population (One place to another place)
— Govt. of India Policy
— Cyclone
— Flood
— Landslide
— Land Degradation
— Drought
2
Restructuring
— Road
— Cannal
— Dam
— Railways
— Infrastructure Development
3
Expansion
— Population Growth
— City Growth (NCR)
— Degradation of Land
— Degradation of Environment Pollution etc.
— Warming of Earth
4
Intensification
Govt. of India Intensified Schemes
— Forest Management
— Watershed Management
— Land use Development Plan
— Agricultural Development
— Socio-Economic Development
— To Arrest Terrorism
5
Sustainability
India acts to restore food grain to sustain the growth of population in future.

41

Food Security and Geospatail Analysis
Some Comments

Don Funnell

Food Security and Geospatial Analysis -I

1. In recent years the food security debate has focused on:

   - Recognising that the link between food availability and nutritional well-being is not automatic.
   - The impact of new agendas such at globalization, biotechnology etc.
   - Questioning whether food security remains a useful policy concept.
   - Querying the role of food aid (ODI briefing March 2002).

2. At the same time there has been a rapid growth in the development of affordable technologies associated with the provision of data which are flagged to assist in the decision making process. Information and communication technologies (ICT) have been developed which promises not just improved methods of achieving existing aims but, arguably more important, providing the capacity for isolated and disenfranchised communities to participate in the whole process of managing food security in a manner that offers new ways of thinking about the process (Deichmann and Wood, 2001).
3. Thus we are faced with examining suggestions that not only

is the food security debate becoming more nuanced in the light of a better understanding of the food and nutrition system but that ICT, especially geo-spatial technologies have become increasingly significant in tackling these problems.

Geospatial Analysis

The role of geo-spatial analysis in the Food Security debate became clear once it was recognised that the analysis of FS required *monitoring* the joint process of demand side and supply side factors in the food chain. Whilst there have been many theories about the cause and evolution of food supply problems, all agree that the supply of appropriate information is crucial.

Thus Geospatial analysis must be involved in the following processes :

1. *Generating data* : In this category falls the wide range of information gathering technologies ranging from the 'high tech.' end of satellite imagery and the use of GPS, through structured data collection by formal agencies, for example on the state of food stocks and offcial prices, though to a large range of informal data sets collected often by PRA techniques. For Food Security there remains debate concerning exactly what information is most desirable to collect and how frequently. The single unifying feature of such data sometimes its most elusive quality other than reliability.... that it has to be spatially referenced. For geospatial analysis such information must effectively be assigned ȧ specific (unique) location.
2. *Handling Data :* Under this heading we need to note the particular forms in which data are categorised and stored. In the early days of FS policies most of the data was in handwritten form. Today an increasing amount is in digital form. However, the 'mixed' form *of* information, the problems of transferability, and more recently the incompatibility between different digital storage systems remain barriers to effective use. Different agencies involved in the FS scenario use different data bases!

3. *Interpreting data* : Food security policy usually requires the monitoring of the flows of food throughout an agricultural year. Depending upon the precise configuration of the supply chain will depend upon the focus of attention concerning vulnerabilities. If a particular area (country) is highly dependent upon imports then global supply prices are critical. By contrast domestic production capabilities may be the crucial and the classic 'early warning scenarios using various methods of estimating harvest volumes provide the focus of attention (The most common approach using satellite imagery). In some cases food vulnerability is *not* a function of gross supply or even of price *per se* but of weaknesses in other aspects of infrastructure such as transport, especially in the face of natural disasters such as earthquakes or floods.
4. *Data interpretation :* Therefore is a particularly critical element in the process of using geospatial information. I shall return to this issue in the second lecture.
5. *Developing Policy* :In terms of geospatial analysis two important points arise. First that it is important to ensure that the domain of policy for FS itself takes account of the geospatial frame of reference of the available data. National, regional, and local scales of policy formulation only really make sense when it is possible to obtain and analyse data which itself is available at these scales. Too often, there is a mismatch, largely because of the shortage of spatially referenced data at each of these scales. On the other hand, in many areas there is an increasing focus *not* on areas at all but on individuals. Targeted food aid raises the whole question of whether geographical aggregates (spatial units such as local government areas/villages etc.) are the correct framework for policy focus rather than the more precise but difficult arena of individual/family selection. More and more NGOs may be trying to take this approach because of problems with more aggregated targeting.
6. *Monitoring :* This is perhaps the most valuable area in which ICT may play a part. One of the problems with most conventional data gathering procedures concerns the infrequency of data availability. Traditional annual crop

returns, beloved of Ministry of Agricultures etc. can take months to process. Using Rapid Appraisal techniques combined with the input from satellite monitoring may help to update information continuously and spot anomalies that can then be-checked.

These key procedures form the basis of any Food Security Section. They are, however, subject to many broader questions concerning the institutional structure in which they can be successfully placed. I will deal with these questions in my second presentation and turn now to looking directly at some examples of current GIS applications in the Food Security field.

GIS, Remote Sensing and the Question of Vulnerability

I have chosen to focus discussion on this issue because it seems to me to link the technical requirements of GIS systems (spatial referability, statistical validity etc.) with the more policy orientated factors concerning the food security operations and policy.

Some definitions of vulnerability include

- Defencelessness, insecurity, and exposure to risk, shocks and stress and difficulty in coping with them Vulnerability thus has two sides : an external side of risks, shocks and stress to which and individual or household is subject and an internal side which is defenselessness, meaning a lack of means to cope without damaging loss (Chambers, 1989);
- The degree to which a system may react adversely to the occurrence of a hazardous event (Timmerman, 1981);
- Vulnerability is an aggregate measure, for a given population or region, of the underlying factors that influence exposure to famine and predisposition to the consequence of famine (Downing 1991);
- Vulnerability is used to describe the relative susceptibility of households to various levels of food insecurity (FEWS, 1993). *Source :* (Ramachandran and Eastman, 1996)

Their study highlights the wide range of approaches used to measure vulnerability and therefore to provide appropriate data into a GIS.

Some example include (Zambian data) :

- *Measures of Crop Risk* : which is a function of the average length of the growing season arid the share in total output (agric) of drought resistant crops.
- *Market Access :* which is a function of average cost of travel to nearest district market and average cost of travel to nearest major urban market.
- *Coping Strategies :* this includes, per capita livestock units, per capita fisheries production, staple food production per capita, percentage of female headed households (more vulnerable), average percentage of households with less than one month's reserve.

These data are combined to form it *vulnerability index,* which in turn is mapped to identify zones of particular potential difficulty. This example is relatively simple. By contrast, in the case of Ethiopia a IFPRI study has attempted to utilise 77 odd variables to derive a satisfactory quantitative measure of vulnerability. The Ramachandran and Eastman study shows that each region tends to adopt its own sets of criteria hour which Food Security decisions are then derived.

However, the more sophisticated the analysis the greater 'sensitivity' of the findings to misinterpretation and the loss of individuality of the original data. Using composite scores for instance, whilst easily absorbed into GIS operations may fail to distinguish different causal patterns : one area experiencing food stress due to failure of the rains, another because of its excessive reliability on sales of a cash crop from which food is then purchased.

Using Fuzzy Ionic

One development attempts to use 'Fuzzy set' scaling, a technique well known in the statistical analysis of indeterminate values. Using the FUZZY module of IDRISI (and equivalent software packages) helps in the construction of data sets in which individual items are assigned to groups by a 'soft' scaling of values. Natural phenomena, like the fragility of the environment, rarely have crisp boundaries and therefore cannot be approached to the classical way with binary logic (fragile or not : 0 or 1). With fuzz, logic, entities are allowed to have a partial membership to a class, which is assigned by membership functions. With fuzzy relational calculus a mathematical basis is constructed to deal with all kinds of problems where the "fuzziness" of natural environment cannot be neglected.

More details of this technique are provided in a study of West Africa and the way information from remotely sensed information (usually NDVI type values) is combined with a wealth of socio-economic data is worth further examination (see Ramanchandran and Eastman,1996).

In terms of the technical development of GIS in the field of Food security probably one of the most well advanced systems is that provided by FEWS (Famine Early Warning Systems). Net established by USAID (FEWS, 2000). The point about this project is that it uses the internet to bring together information from the analysis of satellite data (climate, weather, soil moisture, crop health and harvest estimates), economic databases, (crop stocks, prices, transport capacities etc.) and reports (local officer monthly bulletins noting food security issues). It also provides some degree of interactive skill training in the use of these sources that has an important bearing on the types of users of the system (see below). Altogether, although it has its critics and some technical weaknesses (McGuire, 2000) it remains one of the key initiatives in the integration of data sets for food security (Another approach is provided by the FAO) (GIEWS)

Public Participation GIS

It was noted above that the FEWS initiative, through its interactive elements and the use of the internet, aims to broaden the participation of those involved in Food Security Issues. Arguably, this is one of the key issues surrounding the development of GIS in this field Generically, since the early 1980s the political economy of development has moved substantially in the direction of wider participation and Food Security debates have not been isolated from this trend However, until very recently most of the GIS/RS technologies have been firmly in the hands of the national and international agencies who have tended to control the flow of information relevant to managing food security issues.

However, three developments have together somewhat altered this situation :

- Available computing power at (for many) at least NGO prices can use basis GIS and image processing software.
- Cheap and (since the removal of selective availability) relatively accurate GPS equipment enable sources of data (households etc.) to be accurately geo referenced with ease.
- More and more people and agencies have access to the internet which can act as a coordinating framework both for the different sets of data and sometimes offering online GIS and RS processing capacity. For instance the relatively recent MODIS instrument provides very useful data for land evaluation that can be accessed for preliminary viewing via the web. Together these developments allow the growth to GIS capacity amongst a much wider clientele, and serve as the technical basis for Public Participation GIS (PPGIS).
- The development of PPGIS (sometimes known as GIS2) has proceeded apace not least in the field Food Security In addition to the direct technical issues tire political message associated with the development should also be recognised. However, it is important to set out some of the key issues facing PPGIS and see how they relate to the Food Security debates.

The Table 41.1 establishes five main sets of criteria is to be considered in the construction of PPGIS and we can briefly examine these in the light of the conventional framework under which GIS is used in a Food Security Framework. Under 'communications structures for instance most existing systems fall somewhere between 'Independent Nodes and a Radial Framework', reflecting the organisational structure of the current group of stakeholders, most of whom are official agencies or NGOs. To some extent the FEWS system mentioned earlier approaches the web based framework. An examination of much of the writing on GIS and FS shows that there remains very little discussion on the issues associated with the legal and ethical dimensions of data and analysis for Food Security, although thus is now gradually entering the debate.

Thus whilst PPGLS may be the direction of much current research

Table 41.1 : Different Models of PPGIS Availability

Dimensions	*Attributes*
Communication Structures	Independent nodes Radial Connectivity Network Connectivity
Nature of Interaction with GIS	No Direct Use Passive Use Active Use Proactive Use
Location	In House GIS Virtual (Web based GIS) Remote GIS (outside community)
Stakeholders	Local and Non-local State Agencies Non-Government Organisations (NGOs) Educational Institutions Private Sector Institutions; Within Community Stakeholders
Legal and Ethical Issues	Ownership of/responsibility for spatial data bases Access to publicly held information Checks and balances governing appropriate GIS use

Source : Ieitner *et al.* (1998).

activity, alongside technical advances such as the integration of narrative data sets, multimedia GIS and non-hierarchical systems of information management, little of these advances have permeated the mainstream Food Security applications. Put bluntly, although research is needed to refine the technical use of many of the data sources, especially in the remote sensing field and in improving data reliability (especially when geo referenced), much of the future emphasis will undoubtedly focus on the institutional framework in which GIS can be successfully employed m advance our management of food security issues. This is the topic of my second lecture.

Food Security and Geospatial Analysis - II

1. In the last lecture 1 pointed to the increasing interest by the GIS community to promote the development not just of new

technical advances but also with the construction of PPGIS Although the feasibility of such an approach is partially determined by technical developments the impetus comes quite clearly from the increasing recognition that the effective application of GIS depends as much, if not more on the institutional framework in which it is employed. In relation to food security note the following :

> *... the principal problem is no longer nor knowing why, when, or where food crises are threatening: it is ensuring that the information is used to best advantage, to elicit a timely and appropriate response.* (Buchanm-Smith, 1997)

This has become one of the leading questions in the wider debate about the particular initiatives appropriate far agencies at all spatial scales (see ODI briefing).

2. There are many complex issues that influence the structure of FS institutions, and the way in which GIS *related data* collection and analyses feeds into a 'decision support system'. Some very general issues are examined for India in a thesis by Bank (1997) and in a recent paper by Mehta (2000).

Danik's thesis provides a detailed enquiry into the relative effectiveness of the Indian food security framework that has, by and large, managed to contain widespread serious famine. However every year some region of India suffers from a food crisis and since the work of Sen the focus has moved away from the simple supply side models to a more comprehensive appreciation of institutional provision Nonetheless, many food crises are initiated by drought and Danik's work explores the extent to which key components, such as an effective early warning service, a responsive administration and most controversially an effective media and political opposition together contribute to the management of food deficiencies. The inclusion of the later topic may seem odd in a discussion on GIS but it is central to the contemporary debate about how early warning systems may operate.

Hints of this are also given in the paper by Mehta who discusses agricultural statistics and identities crop forecasting as a weak link in the Indian system. He suggests that remote sensing can help to improve this. However, the point I would like to emphasise from his paper is that although statistical analysis from RS data produces information with relatively little 'human bias', the role of the 'stakeholders' in the process remains very significant. Levels of accuracy and timeliness of data are not, in practice free from 'negotiation' and it is the way in which the data are fed into and employed by the different agencies that plays a critical role in the adequacy of early warning systems Thus our attention is focused on the role of early warning systems in which GIS plays a major role.

Typology of Early Warning Systems

A famine early warning system has been described by Davies *et al.* (1991) as a system of data collection to monitor peoples access to food, in order to provide timely notice when a food crisis threatens and thus to elicit an appropriate response. In the light of the previous discussions and the focus of this workshop we need to examine just how users employ the results of FWS.

Table 41.2 : Typology of EWSs

	Conventional Famine Early Warning System	*Alternative Food Information System*
Scope	Famttre-oriented	Food Security oriented
Determinants of food security	Food production	Access to food
Level of operation	Macro centralized	Micro decentralized
Unit of analysis	Geogrphic, e.g. nation/districts	Socio-economic, e.g. vulnerable groups
Approach	Top-down Data-centred	Bottom-up People-centred
Response	Food-aid oriented	Sustainable improvement in accès to food

Source : Davies *et al.*, 1991.

You will see immediately that whilst the first type (famine of oriented) dominates the current practice more and more agencies are moving to develop the broader Food Security approach. Most local level systems tend to be of the 'food security type' because they are able to handle more complex data generated by individuals and groups with direct input into the food supply chain. At the macro scale, although the data may be highly technical (satellite images etc.) its information content is highly simplistic to that it offers only an indication of probable land cover/crop state often at a rather crude geographical scale. In turn this offers considerable scope fore some of the most powerful uses of GIS in decision systems because of the ability to compile and monitor stocks and flows of resources in relation to the geographical patterns of demand. However, as the demands for geographical targeting increase (see Biman and FoFack, 2000) so then do the requirements for geographically referenced data. Bottom-up approaches demand a much finer scale of geo-referenced data but for many territories this still remains a long term goal but it will surely come to be an integral part of most Food security systems.

In the shorter term there are still many questions revolving around an understanding of factors affect the efficacy of the EWS data?

The Timing of Response

Whilst much of the technical discussion focuses on the nature of EWS information, for example the extent to which spatially referenced data is available, it is very important to emphasise that user agencies are particularly sensitive to the timing of information. The following diagram, derived from Buchanan Smith and Davies (1995) illustrates this point.

The diagram highlights the fact that the success of the application of the technologies *via* an EWS depends crucially upon the timing of the uptake of the information. Information services, that is better predictive indicators of, for example climate or crop variables *via* Remote Sensing and greater ability to pinpoint geographically localised issues through improved GIS capabilities are major contributions that research into GIS/RS can contribute. However, the efficacy of the system is linked to the factors which determine when agencies are prepared to ACT on the data provided. The evidence from studies of existing FEWS suggest that there are many institutional issues that effectively delay the action from the *timely* (above the box) arena to the *ineffective* (below the box) situation.

A Clear and Consistent Message

Whilst the technical improvements in FEWS through GIS etc. have

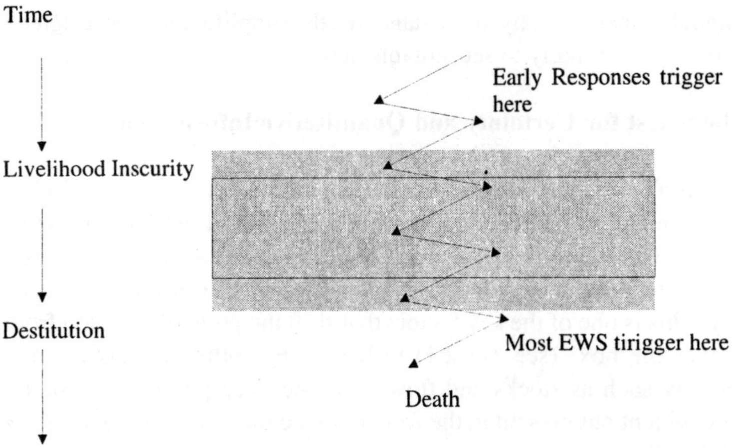

Fig. 41.1 : Food Security Management and the Timing of EWS After Buchanan-Smith and Davies, 1995.

been impressive this does not mean that the message provided to decision maker is necessarily clear and appropriate. As was noted in the previous lecture, the interpretation of the information is still very much an Art, not a Science and no amount of statistical information will necessarily improve this. In some cases the problem arises because there is a duplication of effort as different agencies, national and international seem constitutionally determined to assert the validity of their own interpretations and the validity of 'their data' (see below). This is an interesting comment as it rather flies in the face of the issues of widening access. From one perspective several different sources of data allow full debate on the interpretation of the evidence but there is no doubt that, for many executive agencies of governments and NGOs such complications have led to unnecessary delays.

An interesting study which examined the E1 Nino impacts (Thomson *et al.,* 1998) on potential food security issues highlighted the dilemma in which highly sophisticated probability models of likely drought impacts by locality simply presented a far too confusing picture for most decision agencies. By contrast another approach adopted in some African countrier is based upon the fact that the executive agencies agree on a series of warning 'states' or alert criteria; the FEWS approach simply issues the appropriate level of warning.

Though rather crude by many standards the simplification, once agreed is much more likely to see suitable action.

The Quest for Certainty and Quantitative Information

This is a point made strongly by Buchanan-Smith arid Davies (1995). Because the predictive outputs from a GIS/RS based EWS must be tinged with some degree of probabilities many agencies are often unable to provide the necessary commitments until too late in the day. This is one of the key factors that shift the point of decision from above 'the box' (see Fig 2.1) to below. For other aspects of food security such as stocks and flows data the uncertainties are usually less evident but present in the form of inaccurate (suspect) data. Thus whilst GIS presentations may give the image of highly sophisticated analysis, a belief in the fragility of the data will easily undermine the call for action. There are many examples of this connected also to the overall degree of trust between the various stakeholders.

The Ownership of Data

Many commentators are beginning to direct attention to this institutional variable. Earlier it was noted that the timeliness of action may be related to the degree of 'trust' between agency and information supplier. The picture is, however, more complicated. International donors are often very sceptical of sources produced solely by national bodies simply because of the complex web of political issues which surround food aid and emergency relief. This is an example where greater cooperation is the key to greater efficiency, or more timely delivery, as has been illustrated in the FEWS models. Nonetheless it is not foolproof. The increasing 'technical' approach to some Food Security issues, and that includes the 'construction of GIS based systems do not provide foolproof indicators of food or nutritional problems. They are only as good at their design criteria and as the emphasis swings away from famine reporting (management) to more subtle issues of broader food and nutrition requirements so the rather crude 'indicators' or thresholds may be inadequate. This is true when the food question is packaged with other elements of a genuine crisis such as health/disease problems. An outbreak of a serious epidemic may be the first sign of a food security failure in some areas due to lowered resistance. Great reliance on a highly technical system focused on crop failure would be unable to pick this up. It is in this situation where local media or local based FS warning systems are most appropriate although there is no guarantee that the message will be heard at a higher level.

The Long Term

It is clear that alongside technical improvements, the greatest arena of innovation for effective food security has to come from developing institutional structures that can use Effectively information provided by GIS etc. It is an ongoing issue because, whilst in many countries the devastating famines may be a feature of the past the poverty and fragility of the infrastructure as well as the dynamics of the natural environment make disasters of all kinds no less likely than previously. In these conditions systems to provide timely warning of a crisis and information on the state of the relevant variables (food supplies, emergency aid, transport capacities) are vital components to the decision-making system. But we have to recognise that such as system is highly political in the best of circumstances (see the Foot and Mouth in the UK). Long term, accurate monitoring of the key variables demand resources : these are easily reduced as the 'fear' of a crisis recedes and yet this is precisely the situation where such investment is cost effective. Appropriate agencies need to cooperate and to do this they must have joint use of data and mutual discussions of the form and content of any information provided. GIS and the related technologies can material assist in this process but it calf equally be oversold. If the data is poor or not timely it is more useless than a collection of worn exercise books! If the information is presented in an unintelligible manner it will be distrusted. It is not a matter of convincing specialists but, often in the last resort to providing politicians with suitably clear pictures of the scenarios. They have to make the decisions, not the GIS!

REFERENCES

Banik, D. (1997) *Freedom* from *Famine* Ph.D. submitted to the department of Political Science University of Oslo.

Barndt, M. (1998) Public Participation GIS: Barriers to implementation Cartography and Geographic Information Systems 25 (2) pp. 105-12.

Bigman, D. and Fofack, H. (2000) *Geographical Targeting or Poverty Alleviation.* World Bank Washington DC. N.L. Buchanan-Smith, M. and Davies, S. (1995) *Famine Early Warning* and Responser — the *Missing Link.* Intermediate Technology Publications, London.

Carver, S., Evans, A., Kingston, R and Turton, I. (2001) *Public participation*, GIS, and cyberdemocracy: Evaluating on-line spatial decision support systems *Environment and Planning B: Planning and Design* 28, no. 6 (2001) p. 907-921

Chambers (1989) Editorial introduction: Vulnerability, coping and policy. *Institute of Development Studies Bulletin,* Vol. 20, 21-7.

Clark; M.J. (1998) GIS— Democracy or delusion? *Environment* and *Planning* A 30 (2) : 303-16.

Committee on the Human Dimensions of Global Change (1998) *People and Pixels Linking Remote sensing* and *social Science :* National Research Council Washington DC.

Davits, S.M., Buchanan-Smith, M. and R. Lambert (1991) *Early warning in the Sahel and Horn of Africa : the state of the Art.* Research Report No 20. IDS Brighton.

Deichman U. and Wood, S. (2001) Appropriate technology for sustainable Food Security *IFPRI 1020* Focus 7, Brief 7.

Downing, T.E. 1991 *Assessing socio-economic vulnerability to famine : Frameworks, concepts and application.* Report produced for the US AID/ FEWS Project and Alan Shawn Fenstein World Hunger Programme, Providence, USA.

Leitner, H., McMaster, S., Elwwod, S McMaster, S., and Sheppard, E. (1998) *Models for making GIS available to community organisations : dimensions of difference and appropriateness,* NCGIA Conference Papers.

Mehta, R. (2000) Remote sensing applications in agricultural statistics system: scope, potential and limitations. *Indian Journal of Agricultural Economics* SS(2) pp. 87-95.

Nall, W. and Josserand, H. (1996) The USAID famine early warning system as an information system *Cahiers des Sciences Humaines* 32, no. 4 p. 903-11.

Overseas Development Institute (2002) Food Security briefing Note www.odi.org.uk/keysheets/

Peng, Z.R. (2001) Internet GIS for public participation, *Environment and Planning B: Planning and Design* 28, no.6, p. 889-905.

Pickles, J. (1995)(ed) *Ground Truth: the social implications of GIS.* Guilford. London.

Ramachandran, M. and Eastman, J.R. (1996) Application of GIS to Vulnerability Mapping: a West African Food Security Case Study. *IDRISI GIS' 96. Herausgeber: Institute für Geographic der Universitat Salzburg.*

Thomson, A., Jenden, P. and Clay, E. (1998) *Information Risk and disaster preparedness: responses to the 1997 El Nino event,* Research Report SOS Sahel, London.

Timmerman, P. (1981) *Vulnerability, resilience and the collapse of society.* Environmental Monograph No. I, Institute for Environmental Studies, University of Toronto, Canada.

USAID FE WS Project 1994. *A model for assessing vulnerability to famine in Burkina Faso, Chad, Mali and* Niger. USAID/FEWS unpublished manuscript, Arlington, VA.

42

Emergence of Statistical Scaling in Floods on Channel Networks from Complex Runoff Dynamics

Vijay K. Gupta

Introduction

Physical processes governing floods, namely rainfall, evapotranspiration (ET), runoff generation from hillslopes, and water transport on a nested drainage network, are highly variable in space and time. Spatial variability in physical parameters describing these processes arises due to differences in geometric, hydraulic and biophysical properties among individual channel links and hillslopes in a drainage basin. The number of hillslopes and links increase with the area of a drainage basin representing spatial scale. Therefore, the number of different values that the physical parameters can take also increase with scale. We define this as the number of degrees of freedom (NDOF), because the complexity of spatial interactions among physical processes in a drainage basin can be compared to microscopic molecular interactions in a statistical-mechanical system, which contains a very large NDOF. In order to solve the physical equations either analytically or numerically, statistical and other simplifying assumptions are necessary to reduce the NDOF. However, most of these assumptions cannot be tested directly against empirical observations because the physical parameters cannot be measured. Remote sensing can provide some of the spatially variable parametric information, especially as it pertains to channel network geometry, precipitation and ET. Still, most of the parametric information regarding processes of runoff generation and transport that is required is not observable.

In contrast to the space-time complexity that physical processes producing floods exhibit at the scales of hillslopes and links, empirical observations show that the aggregated behaviour of peak flow statistics exhibits scale invariance or scaling at successively larger spatial scales of a drainage network. Scaling relationships are represented by log-log linear equations between the mean and higher order finite statistical moments of peak flows and drainage areas. The slopes and the intercepts of log-log linear relationships are called the 'scaling parameters'. *Statistical scaling is an emergent* network model called a Peano basin, which has the empirically observed property of mean self-similarity in its branching structure.[8, 18, 19, 21] Several studies since then [2-7, 45] have contributed to a deeper physical understanding of these parameters. Likewise, new empirical studies on nested basins on the Walnut Gulch basin, AZ[10, 23] and the Goodwin Creek basin, MS [9, 24] have prodivded important new empirical insights into the structure of the flood scaling parameters. These physical insights in conjunction with empirical results on the two nested basins furnish a rigorous scientific framework to formulate and test new physical hypotheses using analytical and numerical approaches. This is the key issue that we wish to explain and illustrate in the rest of this paper. Furey and Gupta[25] introduced a physical statistical scaling framework for low-flows, which is similar to the framework for floods.

Definitions of Three Empirical Scaling Parameters

We define three scaling parameters using empirical flood analyses on Walnut Gulch basin, AZ[23] and Goodwin Creek basin, MS [9, 24] which exhibit scale invariance or scaling. The US Department of Agriculture (USDA) has operated both of these basins as experimental facilities. The Walnut Gulch basin is in the semi-arid southwest United States (US) and the Goodwin Creek basin is in the humid, southeast US. Goodrich *et al.* [23] plotted 2-year and 100-year floods against drainage areas using annual peak flow data from the 18 nested sub-basins of the Walnut Gulch basin shown in Fig. 42.1. In the context of the scaling theory, log-log linearity between flood quantiles and drainage area is a signature of statistical scaling [26, 27] which is necessary for defining scaling parameters.

The slopes of 100- and 2-year floods are almost identical to each other on either side of the break. If the slopes of the log-log linear relationship are nearly the same for floods of different return periods, only one scaling exponent describes the aggregation of peak flows across different scales in a scale-invariant manner. Such processes are said to exhibit *statistical simple scaling* [2]. In Fig. 42.1, let θ_1 denote the scaling exponent to the left of break, and θ_2 to its right. These are the first two scaling parameters that we will discuss here. Both of these parameters describe the scaling statistics of the mean and higher order finite moments of peak flows [2]. In Fig. 42.1, the parameter $\theta_1 \approx 0.9$, which is much larger than $\theta_2 \approx 0.58$ on the right hand side. The third parameter is a *critical area, A_c* at which the scaling exponent changes from θ_1 to θ_2. It is around 1 km^2 in Fig. 42.1. The precise location of scaling break cannot be determined from this data set as there is only one stream flow gage between 0.1 and 2 km^2 where the scaling break occurs. The three scaling parameters, θ_1, θ_2 A_c, are examples, which describe the macroscopic or the aggregate behaviour of different components of physical processes that produce floods.

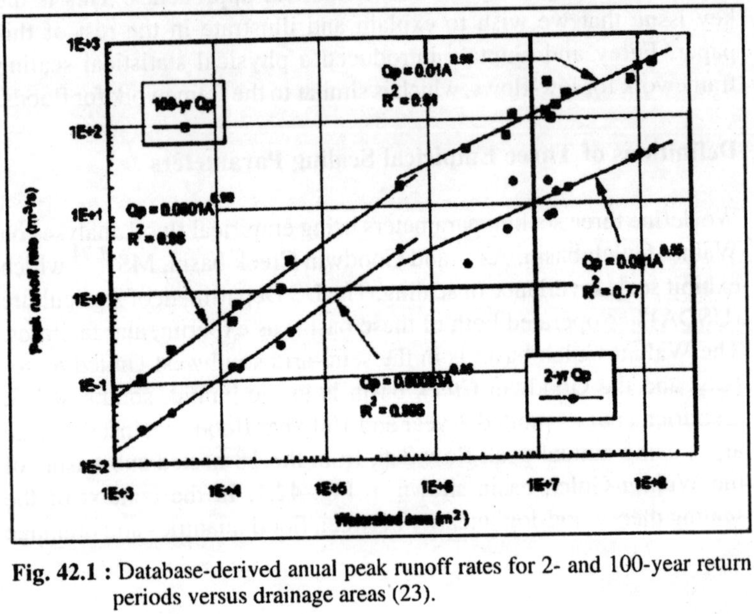

Fig. 42.1 : Database-derived anual peak runoff rates for 2- and 100-year return periods versus drainage areas (23).

Another concept of scaling invariance called *Statistical Multiscaling* has been observed in peak flow data that needs mentioning here for the sake of completeness.[1, 2, 28, 29] In multiscaling processes, multiple slopes and intercepts are needed to describe the scaling of mean and highter order finite moments. They are not multiples of a single scaling exponent as is the case with simple scaling processes shown in Fig. 42.1. Most of the previous work on multiscaling has been published in the context of regional flood frequency analyses.[26, 28, 29] Many studies have investigated the presence of multiscaling in annual flood statistics based on the observation that the coefficient of variation (CV) of annual peak flows does not remain constant as drainage area changes.[27-30] By contrast, if floods obey simple scaling as exhibited in Fig. 42.1, the CV remains a constant with respect to variation in drainage area. We do not discuss multiscaling in this paper; see[1] for a theoretcial example of how multiscaling arises in floods on branched river networks, and[6] for a generalization of those results. Multiscaling has been observed in peak flows from individual rainfall-runoff events on the Goodwin Creek basin.[9]

Tests of Physical Hypotheses Regarding Flood Scaling Parameters

A physical interpretation of scaling parameters requires the context of individual rainfall-runoff events rather than of the annual flood statistics as shown in Fig. 42.1. The tradition of analyzing annual flood statistics was established over 50 years ago in the context of regional flood frequency analyses. By contrast, an event-based statistical scaling analysis is a new appraoch. New, event-based empirical flood scaling analyses have been conducted on the Goodwin Creek basin, MS,[9,24] and on the Walnut Gulch basin. [10] They give a much deeper empirical understanding of scaling parameters compared to annual peak flow statistics shown in Fig. 42.1. For example, both the above studies on the Goodwin Creek basin have shown that the flood-scaling exponent θ_1 for individual rainfall-runoff events varies between 0.9 and 0.6 with a mean of about 0.8. By contrast, this type of an event-based variability in θ_1 cannot be inferred from annual flood quantile analysis shown in Fig. 42.1. Moreover, neither of these two event-based studies found a scaling break within the Goodwin

Creek basin. It implies that $A_c > 21$ km^2. This is much larger than $A_c \approx$ 1 km^2 for the Walnut Gulch basin. This observation suggests the hypothesis that A_c in humid climates is larger than in arid and semi-arid climates, which has been empirically observed by.[29] Let us discuss how the empirical values of θ_1, θ_2, and A_c can be understood physically in an event-based context.

Let τ denote the set of links constituting a drainage network, and each such link itself represents the outlet of a sub-network. Next, let $q(e, t), e \in \tau, t \geq 0$ be a space-time field representing river discharge, or the volumetric flow per unit time, at the bottom of a link e at sometime $t > 0$. Consider a small time interval of unit length $\Delta = 1$, and let $cR(e, t)$ denote the volume of runoff into a link e from adjacent hills in time interval $(t - 1, t)$, where c denotes the mean area of hills, and $R(e, t)$ is the runoff intensity measured in the units of length per unit time. We are neglecting variations among hill areas to simplify notation, and without sacrificing much of physical generality. Finally, let $S(e, t)$ denote the total volume of runoff stored in a link e in time interval $(0, t)$, and $\Delta S(e, t)$ the change in the total volume of runoff stored in time interval $(t - 1, t)$. Then a discrete equation of continuity for a network-hills system can be written as.[2, 3, 7]

$$\frac{\Delta S(e,t)}{\Delta t} = -q(e,t) + \sum_{f:f=\bar{e}} q(f, t-1) + R(e,t)c \quad e \in \tau, t \geq 0 \quad (2)$$

The sum on the right hand side of (2) consists of discharges from all those links f whose bottom vertex f is the same as the top vertex $\bar{e}$ of link e. The topology and geometry of a network enters explicitly into the solution of (2) through this term as illustrated below by some examples. A momentum balance equation is required for expressing the link storage, $S(e, t)$, as function of discharge at the outlet of a link, $q(e, t)$. In addition, hydraulic-geometric relationships,[31] and a specification of hillslope runoff, $R(e, t)c$ is required for solving (2).[5, 7, 9] The formulation of the continuity equation does not require that the channel network be binary. However, it assumes that no loops are present in the network.

An understanding of scaling in peak flows from solutions of (2) requires some familiarity with channel network models. For illustration, we will mention three comprehensive channel network models that have been introduced within the last 35 years. The most well known of these three is the random model.[32, 33] It is a statistical model that was developed from a magnitude-based enumeration of channel network configurations. The second model, known as the Tokunaga model, was developed from the property of mean self-similarity in the side tributary structure.[34] This model has drawn attention in the recent literature,[18, 19, 35, 37] because it is able to exhibit some of the empirical features of channel networks that the random model cannot. Peckham [18] gives a clear self-contained explanation of Tokunaga model. The third is a newly introduced broad class of statistical models known as random self-similar networks (RSN) [21]. Both the second and the third models are based on Strahler ordering [36] rather than on magnitude. For a sub-class of RSNs, Tokunaga model can be obtained as a special cse.[21] Moreover, many of the random model predictions can be obtained from RSN, but the underlying mathematical structure of these two models is fundamentally different. Other statistical channel network models have been introduced in the literature,[37, 38] but we will not go into a further discussion of network models in this article.

Idealized analytical solutions of (2) on a network have provided important physical insights into the three statistical scaling parameters [2]. As an illustrative example, consider the solution of (2) under three assumptions : (i) change in storage in each link. $\Delta S(e, t) = 0$. (ii) hillslope runoff into each link, $R(e, t) \equiv R(t)$ is *spatially uniform,* where $R(t)$ is the runoff intensity per unit area, (iii) the velocity of flow through out a network is constant. Under these conditions, the river runoff at the bottom of a link, e, on a network is given by a *convolution,*

$$q(e, t) = a \sum_{s=0}^{t-1} We\,(t-1-s)\,R(s) \quad t \geq 1 \quad e \in \tau, \qquad (3)$$

Here, $We, (x)$, $x \geq 0$ is called the *local geomorphologic width function* of a network for each link e. It is defined as the number of links at a distance x from the bottom of a link e representing an *outlet.*[2, 3]

The local geomorphologic width function varies stochastically as a function of e from one link to the next due to natural spatial variability in the branching and the geometric structure of a channel network. Therefore, runoff field, $q(e, t)$ in (3) represents a space-time random field. The dependence of runoff on spatial scale arises through the indexing parameter, e, denoting a link. To illustrate this important issue, define the spatial scale as the total number of links, $2m(e) - 1$, where $m(e)$ is the link magnitude representing the total number of source links draining into link, e.[33] Therefore, the total drainage area, $A(e) = 2m(e) - 1$ if $c = 1$, which illustrates how runoff field, $q(e, t)$ depends on the multiple spatial scales of sub-basins within a basin.

The representation in (3) is fundamentally different from a deterministic representation of river flows that has been widely used in the unit hydrograph theory and in rainfall-runoff models in engineering hydrology. These representations are valid with respect to a fixed outlet, and thereby a fixed spatial scale. As explained below, Eq. (3) has been used to investigate the statistical scaling structure of peak flows with respect to drainage area representing spatial scale. Such analyses cannot be undertaken within the well-establisehd instantaneous unit hydrograph (IUH) framework, or the geomorphologic instantaneous unit hydrographs (GIUH) theory,[2] or rainfall-runoff models that predict the runoff hydrograph at some fixed outlet rather than throughout a basin. Sivapalan *et al.*[39] have discussed the fundamental issue of difference between linerarity and non-linearity in river runoff with respect to temporal and spatial scales, Eq. (3) represents a linear transformation in time, which is similar to the IUH theory.

Gupta and Waymire[2] considered the mean flow conditioned on magnitude, $e[q(e, t)\,\check{}\,m(e)]$, and analysed the scaling behaviour of the maximum of mean flows using Eq.(3),

$$\max E[q(e,t)\,|\,m(e)\,|\,\max_{t} \sum_{s=0}^{t-1} E[We,(t-1-s)\,|\,m(e)\,|\,aR(s), \quad e \in \tau$$

The expression for the mean width function conditioned on magnitude, $E\,[\{W_e(x)\check{m}(e)]$, was taken from[40] that was derived for the random network model[33] as,

$$\lim_{m \to \infty} \frac{mE\,[W_e(\,mx\,\mu)\,|\,m\,(e) = m]}{(2m-1)\mu} \to \frac{x}{2}\,\exp\,(-x^2/4),\ x>0 \qquad (5)$$

The expression on the right hand side of Eq. (5) is that of a Rayleigh density. Here, μ is the mean link length, and (2*m* – *1*) μ is the total area under the conditional mean width function. Result in Eq, (5) has been generalized and obtained using other analytical methods; see [3] for a review.

Gupta and Waymire[2] used the asymptotic result in Eq. (5) to analyse the scaling structure of peak of mean flows using Eq.[4]. Consider a runoff event of finite duration, *T,* from a hillslope, and assume that runoff intensity is uniform over the duration, $R(t) = R$, $0 \le t \le T$. Under these conditions,[2] analysed how, $\max_t E[q(e, t)\ \check{}m(e)]$ varies with respect to scale, *m*(*e*). The results are shown schematically in Fig. 42.2. One can observe the appearance of a break in scaling due to the effect of runoff duration, *T* on the location of the break. *T* represents rainfall duration in this example, since hillsloope processes of runoff generation from rainfall are omitted. The values of scaling exponents on either side of the break are also given. The patterns of these three parameters can be physically understood as follows.

Small sub-basins reach equilibrium as rainfall persists. This means that the entire drainage basin above each outlet contributes to the peak flow. Therefore, the peak flow becomes proportional to the drainage area with an exponent 1 as shown in Fig. 42.2. Ogden and Dawdy [24] mentioned that this phenomenon occurs more often in nature than one would think, albeit in small areas. The time to equilibrium is a special case of the 'concentration time' at which the river discharge is defined to reach its maximum value. The discharge remains in a steady state as long as $R(t) > 0$ after equilibrium is reached. If $j(T)$ denotes the critical magnitude of the largest sub-basin that has reached equilibrium at a fixed duration *T*, then the peak flow scaling exponent is 1 for all sub-basins with magnitude, $k < j(T)$. The saturation extends to still larger sub-basins as duration *T* increases. It suggests the hypothesis that the critical area, A_c increases with T. Castro [20] observed this feature using numerical simulations on a Peano channel network. Menabde and Sivapalan[5] also noted this feature using numerical solutions of the mass balance equation under more realistic flow routing conditions than considered by.[2, 30]

Let us explain to what degree data supports the above physical hypothesis. The critical basin area in the semi-arid climate of Walnut

Gulch is much smaller than in the humid climate of Goodwin Creek basin as noted above. This implies that the durations of runoff events from hillslopes in Goodwin Creek are larger than in Walnut Gulch. Mesa and Mifflin[41] tested the hypothesis that a slow sub-surface flow component of hillslope response contributes to the peak flow in a sub-basin of Goodwin Creek. In that study, for a rainfall event of 30 min duration, the combined duration of fast overland flow and slow sub-surface flow from a hillslope was estimated to be, $T = 2$ h. By contrast, there is no base flow component in the semi-arid Walnut Gulch,[23] and infiltration further reduces the 'effective duration' of rainfall events that produce runoff. Current work based on numerical solutions of (2) suggests that $T = 10$ min gives $A_c = 1.3$ km^2 [10]. Comprehensive numerical tests of the physical hypothesis regarding A_c are being conducted by our group for the Walnut Gulch and the Goodwin Creek basins. These tests will also consider the effect of partial storm coverage and channel infiltration on A_c for the Walnut Gulch as suggested by,[23] and of spatially variable infiltration and runoff generation.

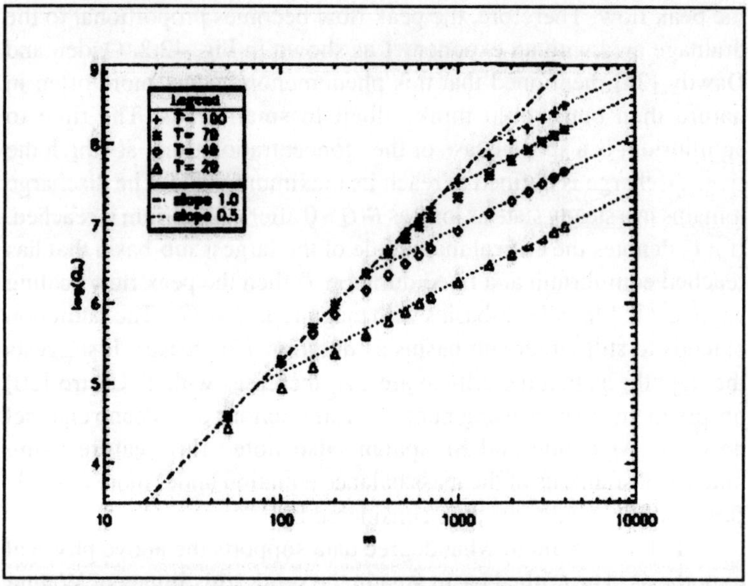

Fig. 42.2 : Schematic depiction of the effects of rainfall duration and basin size on the scaling exponents for floods [2].

Menabde and Sivapalan[5] have used numerical simulations to investigate the patterns of flood scaling exponent θ_1 on the Mandelbrot-Vicsek tree.[42] It exhibits mean self-similarity in its branching structure that is similar to the Peano tree. They simulated finte duration, spatially variable rainfall on the Mandelbrot-Vicsek tree through a random cascade model, which has been very widely used for describing spatial rainfall variability; see a review article by.[43] Results lead to a physical hypothesis that spatial variability in rainfall reduces the value of θ_1, because all the hillslopes do not produce runoff at equilibrium, [5 Fig. 42.6]. Also rainfall advection is observed to reduce this exponent below 1. A new empirical finding has shown that the empirical estimates of θ_1 vary between 0.9 and 0.6 for different rainfall events on Goodwin Creek basin.[9, 24] We are conducting comprehensive tests of this physical hypothesis using numerical simulations on the Walnut Gulch and Goodwin Creek basins. We are also investigating how spatial variability in runoff generation affects θ_1 .[9]

Past work suggests the physical hypothesis that θ_2 is determined by the branching and the geometric structure of channel networks and by the flow dynamics. In addition, spatial rainfall variability also affects it. Let me explain how this hypothesis has developed beginning with the paper by.[1] Fig. 42.2 predicts that the scaling exponent, $\theta_2 = 0.5$ to the right of the critical magnitude or area, A_c. The value 0.5 arises due to scaling of the peak of the mean width function for the random model of channel networks. It holds in the limit of large basin magnitude (or area) as shown in (5). By contrast, the deterministic Peano network model predicts, $\theta_2 = 0.792$.[1,20] It represents the fractal dimension of the maximum contributing set of a peano tree. Therefore, it is also an asymptotic result. Both these network models exhibit mean self-similarity in their branching structures.[18] Veitzer and Gupta [8] studied the scaling of the peak of the width functions for a sub-class of RSNs. They found that, $0.4 < \theta_2 < 0.65$ for 38 different parameterizations of RSNs.

Menabde *et al.*[4] found using numerical simulations that in the presence of dynamic storage, $\Delta S(e, t)/\Delta t > 0$ in (2), scaling holds asymptotically for large drainage areas. This is an important result because it shows the emergence of flood scaling in the presence of non-trivial dynamics on self-similar channel networks, albeit in an asymptotic sense. Moreover, the scaling exponent $\theta_2 = 0.63$ was found to be below its value of 0.792 predicted by the peak of the width function for a Peano network. An analytical demonstration of this result remains open. For RSNs, they observed that the two exponents were very similar, 0.57 for the peak of the width function and 0.53 with dynamic storage. Menabde and Sivapalan[5] found that in a Mandelbrot-Vicsek network the non-linear velocity lowers the scaling exponent θ_2 more than due to dynamic storage alone. By contrast, empirical estimate of θ_2 on the Walnut Gulch basin found it to be greater than the scaling exponent for the peak of the width function. Clearly, this empirical observation is a new finding and it cannot be explained based on the results described above. Current work by[10] suggests using numerical solutions of (2) that this discrepancy may be resolved by relaxing the assumption of uniform Chezy friction on the Walnut Gulch network.

Finally, let us consider how spatial rainfall variability affects θ_2. Gupta *et al.*[1-3] showed analytically for a Peano network using a random cascade model of rainfall that θ_2 decreases below its value predicted by the peak of the width function due to rainfall intermittence. This result has been generalized to more realistic mean self-similar networks than Peano by.[6] Meanabde and Sivapalan[5] (Fig. 42.6) found the same result for finite duration rainfall. We are currently testing this hypothesis on the Walnut Gulch basin.

Closing Remarks

In closing, I will make some brief remarks on how the scaling theory compares with existing rainfall runoff models and regional flood frequency analyses. Both of these approaches are well established in the hydrology literature. In rainfall-runoff models, a simulated hydrograph is compared with observed hydrograph at a fixed spatial scale. While some of the parameters can be related to observable watershed characteristics, most parameters are conceptual representations of non-measurable watershed characteristics that are estimated through a calibration procedure involving subjective statistical procedures. Hydrologists have used calibration of parameters to reduce the enormous complexity of the problem due to a very large NDOF. Because of calibration, rainfall-runoff models are not quite suitable for testing specific physical assumptions and hypotheses within a scientific framework.

By contrast, in the scaling theory, hydrographs are simulated at all junctions throughout a network that is partitioned into links and hillslopes. Much of the physics is reflected in scaling exponents, which provide a measure of rate of change of physical variables with respect to changes in scale. As illustrated here, scaling parameters can be predicted from physical processes and the values can be compared against empirical values for testing different physical hypotheses and statistical assumptions. This is the key new idea, and it has profound implications for the way spatially variable physics of rainfall-runoff processes can be understood within a rigorous scientific framework.

Predictions from ungauged basins have been based on statistical regressions of observed peak flow statistics in a geographic region without any knowledge of physical processes that transform rainfall to runoff. Rainfall-runoff models and regional flood frequency analysis have remained mutually disconnected despite recognition in the literature that physics of rainfall-runoff processes needs to be linked with its statistical variability [44]. The scaling theory tackles this long-standing, fundamental unsolved problem because it unifies spatial flood statistics with the underlying physical processes. As this theory matures and develops further, it would have many applications. For example, it provides a new scientific framework for solving the problem of prediction of floods from ungauged and poorly gauged basins (PUB) globally. The PUB problem includes diagnosing the effects of land use changes due to urbanization and deforestation and due to climate on floods.

Acknowledgements

Gratitude is expressed to Dave Dawdy, Peter Furey, Ricardo Mantilla, Jim Smith, Brent Troutman, and Seth Veitzer, for several useful discussions, which greatly helped in clarifying this presentation. This research was supported by grants from NSF and NASA.

NOTES

1. Gupta V.K., Castro, S., (1996) Over T.M. On scaling exponents of spatial peak flows from rainfall and river network geometry. *J Hydrol*; 187:81-104.
2. Gupta, V.K., Waymire, E. (1998) Spatial variability and scale invariance in hydrologic regionalization. In : Sposito, G., editor, *Scale dependence and scale invariance in hydrology*. London : Cambridge; p. 88-135.
3. Gupta, V.K., Waymire E.C., (1998) Some mathematical aspects of rainfall, LaNDO forms and floods. In : Barndorff-Nielsen, O.E., Gupta V.K., Perez-Abreu, V., Waymire E.C., editors. *Stochastic methods in hydrology : rainfall,* LaNDO Forms and floods, Advanced series on statistical sciences and applied probability, Vol. 7. World Scientific; pp. 129-71.
4. Menabde, M., Veitzer, S., Gupta, V.K., Sivapalan, M. (2001) Tests of Peak flow scaling in simulated self-similar river networks. *Adv. Water Resour*; 24 : 991-9.
5. Menabde, M., Sivapalan, M. (2001) Linking space-time variability of rainfall and runoff fields : a dynamic approach, *Adv Water Resour*; 24:1001-14.
6. Troutman, B., (2001) Over, T.M. River flow mass exponent with fractal channel networks and rainfall. *Adv. Water Resour*; 24:967-89.
7. Reggiani, P., Sivapalan, M., Hassanizadeh, S.M., Gray, W.G. (2001) Coupled equations for mass and momentum balance in a stream network : Theoretical derivation and computational experiments. *Proc Roy Soc Lond* A; 457-89.
8. Veitzer. S., Gupta, V., (2001) Statistical self-similarity of width-function maxima with implications for floods. *Adv Water Resour*; 24:955-65.
9. Gupta, V.K., Troutman, B., Dawdy, D., Karlinger, M., bates, B. Tests of statistical multiscaling in peak flows on channel networks. *Water Resour Res,* submitted for publication.
10. Mantilla, R., Gupta, V.K., Mesa, (2003) O. Tests of a hypothesis of flood-scaling exponent in terms of downstream hydraulic-geometric variations in channel friction (preprint).
11. Lienhard, J.H. (1964) A statistical-mechanical prediction of the dimensionless unit hydrograph. *J Geophys Res*;69(24) : 5231-38.
12. Gupta, V.K., (1983) Waymire. On the formulation of an analytical approach to hydrologic response and similarity at the basin scale. *J Hydrol*; 65 : 95-123.
13. Martin-Lof, A. (1979) Statistical mechanics and the foundations of thermodynamics. *Lecture Notes Phys.* 101.
14. Turcotte, D.L., Rundle, J.B., (2002) editors. Self-organized complexity in the physical, biological and social sciences. *Proc Natl Acad Sci,* USA; 99 (Suppl 1.); 2463-65.
15. Band, L. (1986) Topographic partition of watersheds with digital elevation models. *Water Resour Res.*; 22(3) : 15-24.
16. Chen, Z.Q., Govindaraju, R.S., Kavvas, M.L. (1994) Spatial averaging of unsaturated flow equation under infiltration conditions over areally heterogeneous conditions 1 and 2. *Water Resour Res*; 30 (2) : 523-35.

17. Duffy, C.A. (1995) Two state integral balance model for soil moisture and groundwater dynamics in complex terrain. *Water Resour Res*; 32(8) : 2421-34.
18. Peckham, S. (1995) New results for self-similar trees with applications to river networks. *Water Resour Res*; 31 (4) : 1023-29.
19. Dodds, P.S., Rothman, D. (1999) A unified view of scaling laws for river networks. *Phys Rev* E; 59 : 4865-77.
20. Castro, S. (1998) *Scaling exponents of floods from scale-invariant rainfall and river networks.* M.S. Thesis, University of Colorado, Boudler, CO, USA.
21. Vietzer, S., Gupta, V., (2000) Random self-similar river networks and derivations of generalired Horton laws in terms of statistical simple scaling. *Water Resour Res*; 36(4) : 1033-48.
22. Rodriguez-Iturbe, I., Rinaldo, A. (1997) *Fractal river basins.* Cambridge University Press.
23. Goodrich, D., Lane, L.J., Shillslopeito, R.M., Miler, S.N., Syed, K.H., Woolhiser, D.A. (1997) Linearity of basin response as a function of scale in a semi-arid watershed. *Water Resour Res,* ; 33(12) : 2951-65.
24. Ogden, F., Dawdy, D.R., Peak discharge scaling in a small Hortonian watershed. *J Hydrol Eng,* ASCE, in press.
25. Furey, P., Gupta, V.K. (2000) Space-time variability of low stream flows in river networks. *Water Resour Res,* 36(9): 2679-90.
26. Gupta, V.K., Dawdy, D.R. (1995) Physical interpretations of regional variations in the scaling exponents of flood quantiles. *Hydrol Process*; 9(3/4) : 347-61.
27. Gupta, V.K., Mesa, O., Dawdy, D.R. Multiscaling (1994) theory of flood peaks : Regional quantile analysis. *Water Resour Res,* 30(12) : 3405-21.
28. Smith, J. Representation (1992) of basin scale in flood peak distributions. *Water Resour Res*, 28(11) : 2993-9.
29. Cathcart, J. (2001) The effects of scale and storm severity on the linearity of watershed response revealed through the regional L-moment analysis of annual peak flows. Ph.D. dissertation, University of British Colombia, Vancouver, Canada.
30. Bloschl, G., Sivapalan, M. (1997) Process controls on regional flood frequency; Coefficient of variation and flood frequency. *Water Resour Res,* 33 (12) : 2967-80.
31. Leopold, L.B., Wolman, M.G., Miller, J.P. (1994) *Fluvial processes in geomorphology*. San Francisco : W.H. Freeman.
32. Shreve, R. (1966) Statistical law of stream numbers. *J. Geol,* 74 : 17-37.
33. Shreve, R.L. Infinite topologically random channel networks. *J. Geol,* 1967; 75 : 178-86.
34. Tokunaga, E. (1966) The composition of drainage network in Toyohira River basin and valuation of Horton's first law. *Geophys Bull Hokkaido Univ,* 15 : 1-19 [in Japanese with English summary].
35. Turcotte, D. (1997) *Fractals and chaos in geology and geophysics*. 2nd ed. Cambridge.
36. Strahler, A.N. (1964) Quantitative geomorphology of drainage basins and channel networks. In : Chow, V.T., editor. *Handbook of Hydrology*. McGraw-Hill Co., pp. 4-39-76.

37. Cui, G., Williams, B., Kuczera, G. (1999) A stochastic Tokunaga model for stream networks. *Water Resour Res,* 35 : 3139-47.
38. Daya Sagar, B.S., Srinivas, D., Prakasha Rao, B.S. (2001) Fractal skeletal based channel network in a triangular initiator basin. *Fractals*; 9(4) : 429-37.
39. Sivapalan, M., Jothityangakoon, C., Menabde, M. (2002) Linearity and non-linearity of basin response as a function of scale : Discussion of alternative defintions. *Water Resour Res.*, 38 (2) : 4-1-5.
40. Troutman, B., Karlinger, M. (1984) On the expected width function of topologically random channel net works. *J Appl Prob*. 22 : 836-49.
41. Mesa, O.J., Mifflin, E. (1986) On the relative role of hillslope and network geometry on basin response. In : Gupta, V.K., Rodriguez-Iturbe, I., Wood, E., editors. *Scale problems in hydrology*. Dordrecht : D. Reidel.
42. Mandelbrot, B., Vicsek, T. (1989) Directed recursive models for fractal growth. *J. Phys A : Math Gen*; 22 : L377-83.
43. Foufoula-Georgiou E. (1998) On scaling theories of space-time rainfall. Some recent results and open problems. In : Barndorff-Neilsen, O.E., Gupta, V.K., Perez-Abreu, V., Waymire, E.C., editors. Stochastic methods in hydrology : rainfall, laNDOForms and floods, *Advanced series on statistical sciences and applied probalility*, Vol. 7, World Scientific, pp. 25-72.
44. Potter, K. (1987) Research on flood frequency analysis : 1983-1986. *Rev Geophys*; 25 (2) : 113-18.
45. Richards-Pecou, B. (2003) Scale invariance analysis of channel network width function and possible implications for flood behaviour *Hydrol Sci J.*; 47 (3).

43

Analysis of Landforms, Soils and Land use/ Land Cover Relationship in Metamorphic Landscape using Remote Sensing and GIS

G.P. Obi Reddy, A.K. Maji and *K.S. Gajbhiye*

Introduction

There is an increasing demand, on the precisely described and classified data on landforms, soils and land use/land cover patterns in order to identify sustainable land use systems that are best suited for the defined geo-ecosystems. Detail information on landform units and their processes in an area is very useful in evaluation of soil resources and land use/land cover patterns for environmental planning (Demek, 1982). River catchments are the most effective physical units for natural resources evaluation, monitoring, planning and environmental management (Cook and Doornkamp, 1974). Geomorphological features are manifestations of underlaying parent material and the nature and duration of geomorphic processes that have produced the associated geomorphic units (Wright, 1993). Landforms mapping involves the identification and characterization of the fundamental units, which compose of the landscape (Fairbridge, 1968). The soils always developed in an organized manner on any given landform and a very close association exists between the different slope units of the landforms and soils (Dent and Young, 1981). Inter-relationship between spatial and temporal sequences is generally studied by establishing soil-geomorphic or pedo-geomorphic relationship (Birkeland, 1990). In order to use land optimally, it is not only necessary to have the information on existing landforms and soils but also on the land use/land cover patterns.

Remote sensing techniques have become the most efficient tools for landform studies and their mapping because of its synoptic view, multi-spectral, multi-temporal capabilities. High resolution satellite data provides reliable source of information to delineate and generate comprehensive and detail inventory of landform units in an area (Reddy *et al.*, 2002). The landforms mapping of a terrain and analysis of their processes helps to understand soils for their precise mapping. High spatial resolution remote sensing data provides valuable information on location, spatial distribution and extent of land use/ land cover. With increase in the spatial, spectral and temporal resolutions, the satellite data becomes effective source for accurate land use/land cover mapping. In the present study, Tundiya river catchment in northern part of Nagpur district has been selected for mapping and analysis of landforms, soils and land use/land cover patterns using remotely sensed data and GIS applications to understand their interrelationships.

Study area

The study area is located between 21° 30′ 00″ to 21°40′ 30″ North latitudes and 79°14′ 00″ to 79°31′ 00″ East longitudes and is located in Metamorphic Plateau in north-eastern part of Nagpur district, Maharashtra, with an area of about 362.18 km^2. The climate is mainly hot sub-humid type with mean annual temperature of 27°C and mean annual rainfall of 1120mm. The elevation of the area varies from 320-580 mt above mean sea level. The natural vegetation of the study area mainly comprises of dry deciduous tree species.

Methodology

The IRS-ID LISS-III satellite data of November 1999 and March 2000 pertaining to the study area was collected and registered to SOI topographical sheets at 1:50,000 scale in EASI/PACE image analysis system. The surface lithological analysis of the study area has been carried out through visual interpretation of the satellite data considering the geological map (GSI, 1978) as a base. The slope map of the area was generated from the Digital Elevation Model (DEM) using the standard in built algorithm of SPANS GIS. Visual interpretation techniques have been followed in delineation of landform units based on the tone, texture, shape, drainage pattern and colour characteristics of the satellite imagery in conjunction with collateral information. Adequate field checks were conducted for deriving information on geomorphic units and to establish the relationship between the image elements and geomorphic unit characteristics. The information on soils was obtained from the available soil survey data of Nagpur district (Anonymous, 1990). The satellite data of November, 1999 and March 2000 was used for land use/land cover analysis. The land use/land cover maps were generated for *kharif* and *rabi* seasons of 1999 and 2000 by independent supervised classification using Maximum Likelihood Classification (MLC) algorithm. The raster layers of land use/land cover map of *kharif* and *rabi* seasons were imported into GIS environment to generate final land use/land cover map. The landform and soil layers have been superimposed on land use/land cover map to extract the land use/land cover properties of different landform and soil units. The information generated on landforms, soils and lands use/land cover from satellite data and collateral data has been analyzed in GIS to ascertain their inter-relationships. Filed checks were carried out in selected sites to verify the boundaries and properties of the delineated landform units and land use/land cover classes.

Results and Discussion

Slope

The nearly level to level slopes (0-1%) are mainly associated with deep valley fills with an area of about 49.50 per cent of the total geographical area (TGA). The very gentle slopes (1-3%) are mostly associated with shallow and deep valley fills and are having 14.30 per cent of the TGA. Gentle slopes (3-5%) are observed in shallow, moderate and deep valley fills and shallow valley fills with an area of about 7.06 per cent. The moderate slope areas (5-10%) are in association with moderately weathered foot slopes and occupy an area of about 10.80 per cent. The majority of the strong slopes (10-15%) are in association with linear ridges and structural hills and are found in the western, north-western and south-eastern parts of the study area covering an area of about 5.62 per cent. Steep slopes (15-30%)

are associated with structural hills and covering an area of 7.56 per cent. The steep slope areas exist mainly in the western, north-western and south-eastern parts of the study area. The very steep slopes are (>30%) mainly spread in the structural hills and linear ridges and covering an area of about 4.30 per cent of TGA. The water bodies occupy about 1.27 per cent of TGA.

Geology

Geologically study area is complex and represented by major formations of Gneiss, Granite and recent alluvium (GSI, 1978). The surface lithological analysis and delineation of different geological units of the study area has been carried out through visual interpretation of the satellite data based on the distinct image characteristics of different formations. Besides the drainage network, rock outcrop exposures and vegetation cover, the geological map of the study area is considered as a base. The prominent rock types that occur in the study area belong to highly metamorphosed Gneiss of Lower Proterozoic Archean age (Krishnan, 1982). The igneous rocks shows dark reddish to gray colour on the imagery and occur as intrusives within the metamorphic rocks. They occupy mainly in northern, central and southern parts of the study area. The major Gneiss formations of the are *Lohangi, formations, Bichuva formation*, *Mansar formation, Charbaol information* belongs to Sausar Group, and *Chlorite Schists formations* belongs to Sakoh Group (GSI, 1978). Granite formations in the form of intrusives occur in small outcrops in southern part of the study area and they show white to blue colour on the image. They are associated with dykes and veins of pegmatite and aplite. This graniticferrous-muscovite-biotite granite is found to have intruded along the planes of the Sausar metasediments. Quaternary alluvium deposits have been identified based on the image characteristics of white to light red colour. They consist of loose, unindurated sequence of interlayered clay/silt and sand beds admixed with cobbles and pebbles. This alluvium is underlain by Gneiss and Lohnagi formations.

Landforms

The underlying lithology, slope and the type of existing drainage pattern have influence the genesis and processes of different geomorphic units. Due to complexity of geomorphic evolution, the catchment shows distinct geomorphic units. The nine distinct geomorphological units (Fig.43.1) have been delineated .and characterized based on visual interpretation of IRS-ID LISS-III satellite data in conjunction with geology, drainage pattern and slope parameters in GIS. *Tabletop summits* are narrow, restricted in nature and are noticed in the different parts of the catchment. These are circular to linear in shape and are influenced by circum-denudation as the dominant geomorphic processes. This unit occupies nearly 3.65 per cent of the TGA. *Structural hills* are noticed as a broad zone of dissected hilly topography in the western part of catchment and occupies nearly 18.95 per cent of TGA. This unit represents a typical stream carved morphology comprising crested radiating eroded ridges along the down slopes. *Subdued plateau* encompasses shallow to medium deposition of basaltic crust and shows entrenchment of stream valleys. This unit is noticed in association with moderate drainage density and stream frequency with multiple slopes. *Linear ridges* are encountered at higher elevations and are formed due to detachment and isolation of various fluvial processes. The weathering and denudational processes in different climatic conditions have detached these relict mounds from the main hill chain and are mostly associated with Gneiss and Lohani formations.

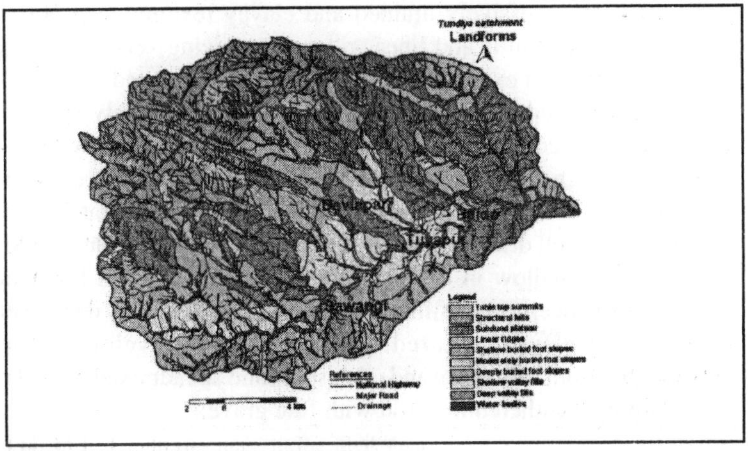

Fig. 43.1 : Landforms maps of the study area.

Shallow weathered foot slopes are moderate in relief and consist of deposited sediments that are regularly carried from uplands of structural and subdued hill slopes. *Moderately withered foot slopes* are moderate in relief and consist of semi-stratified deposited sediments that are regularly carried from uplands areas of the catchments. These are in association with Gneiss, Lohani, Bichua and Mansar formations with sub-parallel drainage pattern and low to moderate density and stream frequency and occupy an area of about 9.52 percent of TGA. *Deeply weathered foot slopes* are low in relief and consist of *in situ* withered material and stratified to semi-stratified deposited sediments. The deposited sediments are admixed with sandy loam and clay loam and occupies with an area of about 14.52 per cent of TGA. *Shallow valley* fills are associated with smooth texture with red colour of image characteristics. These valley fills are unequal in width and are having moderate to steep slopes in valley side floors. They occupy nearly 11.52 per cent of TGA. *Deep valley* fills are narrow slip of low-lying flat zone formed by actions of fluvial deposition on either side of river course and is delineated as deep valley fills. This elongated and narrow zone along with the main channel course consists of moderate to deep soils and it occupies nearly 3.10 per cent of TGA.

Soils

Study indicates that soils on tabletop summits and structural hills are very shallow, excessively drained and clayey textured (Fig. 43.2). Whereas subdued hills and linear ridges are having very shallow to shallow, somewhat excessive drained and clayey textured soils. Soils on shallow buried foot slopes are shallow to moderately shallow, excessive drained and clayey textured. Soils on moderately buried foot slopes are moderately deep to very deep, somewhat excessive to well drained. Soils on deeply buried foot slopes are moderately deep to very deep, well drained and clay in texture. Soils in shallow valley fills are very shallow in uplands and moderately deep in low land areas. Soils in deep valley fills are moderately deep to very deep, well drained and clayey textured. These soils are developed from fluviodepositional alluvium and consist of sand silt admixed with clay underlain by weathered material with fine gravels.

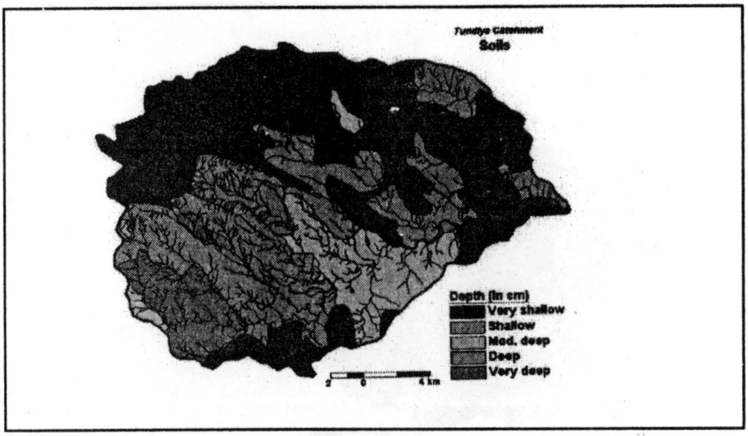

Fig. 43.2 : Soils maps of the study area.

Land use/Land cover

Apart from drainage, slope, landforms and soils of the terrain, the vegetative cover and their spatial distribution is considered to be important parameter in the analysis of landform-soils relationship and the biophysical processes acting there on (Anderson *et al.,* 1976). The analysis of *kharif* and *rabi* seasons satellite data reveals that the structural hills, table top summits, subdued hills and linear ridges are mainly covered with deciduous dry forest and scrub lands (Fig. 43.3). The thick deciduous forest is restricted to parts of structural bills and subdued hills. The shallow, moderate and deeply buried foot slopes are mainly associated with current fallows, scrublands and croplands. Deeply weathered foot slopes are covered with croplands, besides a few pockets of wastelands. The shallow valley fills and deep valley fills under irrigated crops along with sparse vegetation cover. The intensive double-cropped area occupies the main valley floor. Paddy, jowar, sorghum, pigeon pea, pulses and vegetables are the main crops. The water bodies and built up lands are occupied 1.57 per cent of the total geographical area.

Landforms, soils and Land use/Land cover Relationship

The study reveals that landforms mapping and analysis based on the

Fig. 43.3 : Land use/Land cover maps of the study area.

IRS-ID LISS-III data in conjunction with drainage and elevation information is of immense help in better understanding the relationship among landforms, soils and land use/land cover. The analysis also shows that soils on tabletop summits and structural hills are very shallow, excessively drained and clayey textured are associated with dry deciduous forest and scrublands. Whereas, subdued hills and linear ridges are having very shallow to shallow, somewhat excessive drained and clay in texture and are associated with dry deciduous forest and scrublands. Shallow buried foot slopes are associated with shallow to moderately shallow, excessive drained and clayey soils and they are under scrublands, current fallows, and croplands. Soils on moderately buried foot slopes are moderately deep to very deep and are under current fallows and croplands. Soils on deeply buried foot slopes are moderately deep to very deep, well drained and clay in texture and are associated with croplands, current fallows and waste lands. Very shallow soils in upland area of shallow valley fills are in association with scrublands and moderately deep soils in low land areas are associated with croplands. Soils in deep valley fills are moderately deep to very deep, well drained and clayey textured and are associated with cropland areas besides a few pockets of wastelands. Paddy, jowar, sorghum, pigeon pea, pulses and vegetables are growing mainly in moderately deep to very deep soils in moderate to deeply buried foot slopes, shallow and deep valley fills.

Conclusions

The study reveals that integrated remotely sensed data in conjunction with collateral data in GIS enhances the capabilities of interpretation to identify and delineate distinct landform units and land use/land cover classes. The analysis shows that the tabletop summits, structural hills, subdued plateau and linear ridges depict fluvio-denudational geomorphological processes. The shallow, moderate and deep weathered foot slopes, shallow and deep valley fills are shown depositional processes. These fluvial-depositional geomorphic units are formed under the influence of corrosive and erosive dynamics and slope retreat processes besides sediment transportation from upland areas. The analysis reveals that soils on table top summits and structural hills are very shallow, excessively drained and clayey textured are associated with dry deciduous forest and scrublands. Whereas subdued hills and linear ridges are having very shallow to shallow, somewhat excessive drained and clay in texture are also associated with dry deciduous forest and scrublands. Soils in shallow, moderate and deeply buried foot slopes are mostly associated with current fallows, scrublands and croplands. Moderately deep to very deep soils in shallow and deep valley fills are associated with cropland areas besides a few pockets of wastelands. The GIS and remotely sensed data based analysis reveals that drainage morphometry, geology and slope are playing a significant role in genesis and development of landforms and soils. The analysis also shows that the nature of landforms and soils are playing key role in determining the type of land use/land cover existing in the study area. The details obtained in the present study have clearly demonstrated the capabilities of IRS-ID LISS-III data and GIS in delineation and characterization of distinct landform units and land use/land cover classes for analysis of their inter-relationship. The analysis of interrelationship among landforms units, soils and land use/land cover their processes helps in the fields of land evaluation, land use planning, biomass and environmental studies and catchment planning and management.

Acknowledgements

The authors are thankful to Mr. Sunil Meshram, Technical Assistant for his support in GIS analysis and Ms. Rohini Watekar for word processing.

REFERENCES

Anderson, J.R., Hardy, E.E., Roach, J.T. and Wimer, R.E. (1976). *A land use and land cover classification system for use with remote sensing data.* Professional paper, No. 964 USGS. Reston Verginia : 28.

Anonymous, (1990). Soils of Nagpur District, Maharashtra, Report No. 514, NBSS&LUP, Regional Centre, Nagpur. p. 1-68.

Birkeland, P.W. (1990). Soil-Geomorphic research - A Selective overview. *Geomorphology*, 3: pp. 207224.

Cook, R.U. and Doornkamp, J.C. (1974). *Geomorphology* in *Environmental Management: An Introduction*, 413 pp., Clarendon Press, Oxford.

Demek, J. (1982). Geomorphological mapping: progress and problems in *Perspectives in Geomorphology, Vol. IIIN Applied Geomorphology,* (Edited by H.S. Shama), pp. 221-35, New Delhi.

Dent, D. and Young, A. (1981). *Soil Survey and Land Evaluation*. George Allen and Unwsin Ltd., London: 278.

Fairbridge, R.W. (1968). *The Encyclopedia of Geomorphology*, pp 388-03.

Geological Survey of India, (1978). *Geological Quadrangle Map of Topographical Sheet 55 O (Nagpur): Madhya Pradesh-Maharashtra*, Geological Society of India, India.

Krishnan, M.S. (1982). *Geology of India and Burma*, Edison (IV); pp.1-536.

Reddy G.P.O., Maji A.K., Srinivas, C.V. and Velayutham, M. (2002). Geomorphological analysis for inventory of degraded lands in a river basin of basaltic terrain, using remote sensing and GIS, *Jour. of Indian Society of Remote Sensing,* Vol.30 (1 &2), pp.15-31.

Wright, R.L., (1993). Principles in a geomorphological approach to land classification, *Zeitschrif für Geomorphologie*, 16 (4), 351-73, 1972.

44

Developing National Agricultural Market Atlas on WEBGIS

Hema Yadav and *Satyaveer Singh*

Introduction

Agricultural Marketing System involves activities and functions of movement of the produce from producing area to the final consumption. The logistics that are required between production and consumption are of dynamic nature. These logistics of different kinds relating to space and time are real determinants of the marketing costs. The quantification of the spatial movement of the produce from one location to another can improve the whole scenario of agricultural marketing. This is an important activity for efficient, innovative and modern market planning. Thematic presentation of spatial movement and the related data on geographical locations helps understanding the grass root level problems of marketing.

The Govt. of India constituted an Expert Committee on strengthening of agricultural marketing in the country under the chairmanship of Shri Shankarlal Guru. Expert Committee on Agricultural Marketing as well as the sub group of Inter-Ministerial Task Force on Agricultural Marketing, Govt. of India (2002) has proposed Information Technology (IT). Application in agricultural marketing. Sub-group has proposed following major recommendations on I.T.Application in agricultural marketing. These are.

(a) Setting up of value added networks services, Information Kiosks in rural areas;
(b) Promotional E-catalogue for Commodity Profiles;

(c) Farmer advisory services in the fields of agricultural marketing,
(d) National Atlas of Agricultural Markets.

Accordingly, the Institute initiated a pilot research on the project of preparation of Atlas was taken up. The Atlas encompasses various dimensions of data directly or indirectly related to agricultural marketing.

National Atlas of Agricultural Marketing would be "data intensive" and informative model for planning and decision-making. The Atlas is a cartographic and thematic exercise of the quantified data. It has the advantage of presenting the dynamics of Information in visual form. Since "seeing is believing" the visual impact of placement of data on maps have more clarity of the data and leaves long lasting impressions. This all together help dissemination of information in more rememberable and friendly environment.

The "Information Revolution" fueled with high speed, low cost, user-friendly software, helps in removing geographical barriers. Inflow of information, has changed the landscape of Information Technology. Application software like Geographical Information System (GIS) has percolated down to various fields of agricultural marketing, particularly in movement identification of commodities. The GIS technology and Global Positioning System (GPS) with input of real time data from remotely sensed gadgets would be the backbone of the proposed Atlas.

Information for Atlas

Effcient Agricultural Marketing System needs strong backward and forward linkages. The backward linkage means inter-sharing of information in an integrated manner at different stages like, the production, seeds, fertilizers and market. Forward linkages comprise of the information with respect to infrastructure available e.g. market and its facilities, means of transport, storage, processing, pledge-finance, location of the markets, ports, price realization, etc. Hence, the back-end database will have diversity with vast base. Various socio economic variables pertaining to rural areas are also important to be included. Proposed Market Atlas model will have following domains :

- Multi-Database (Socio- Economic, Spatial, Non-Spatial)
- Relational Database (for all secondary data)
- Geomatic Technology (RS, GIS, GPS)
- Distributed System for Internet

Multi-Database : This kind of database requires a systematic and inter-sectoral approach. The Meta database standards are simply a common set of terms and definition that describe the spatial and non-spatial data.

- What data is available?
- From where to find the data?
- How to access the data?

The information needed to create a metadata is often readily available, when the data is collected. A small amount of time invested in making metadata base strong, in the beginning of the project, may save time, resource and material for future. Details of information required for Proposed Atlas are as under:

- Agricultural Production
- Marketable Surplus
- Facilities like Godowns and Cold Storage, Grading, etc.
- Various Systems of Sales
- Demographic Features of Agriculture and Agricultural Marketing

Market-Atlas : is a client server tool designed for Intranet/Internet sites. The system provides thematic presentation of the market information, mapping the logistics and flow of resources, integrated data system for market planning and presentation of data where the inputs could be changed as the marketing environment changes. It is a user friendly GIS front-end to enable multiple local and remote clients to perform routine queries. The software has the capabilities of viewing, query, and spatial analysis, and value added functions on distributed geographic data. The combination of Web Technologies and the power of commercial GIS software enable land use and natural resource managers to analyze GIS data that resides across the Internet. (Fig 44.1 and Fig 44.2)

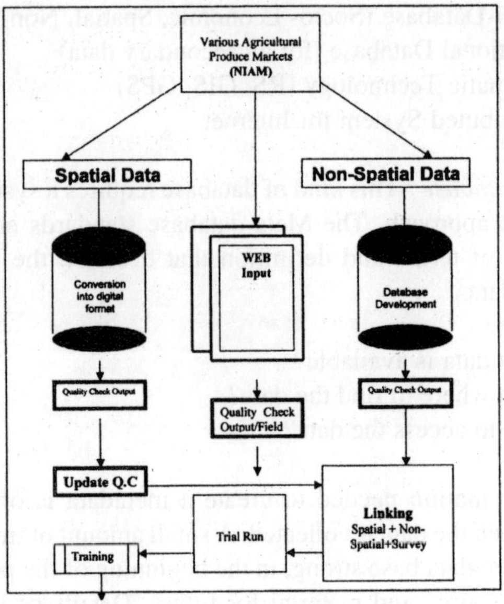

Fig. 44.1 : Integration of Spatial and Non-Spatial Data

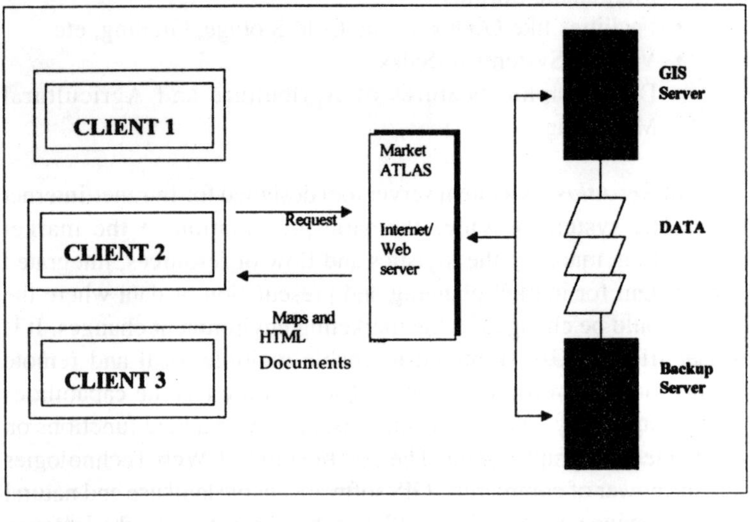

Fig 44.2 : Model for Developing Market Atlas

Model Designing

Requirement analysis involves studying the corrent business system to find out how it works and where improvement should be made. System studies result in an evaluation of how current methods are working and whether adjustments necessary are possible or not. A requirement is feature that must be included in the system. It may include a new way of capturing or processing data, producing information control business activity, or supporing management.

The determination requirement thus entails studying the existing system and details about it to find about what this requirement is.

Activities in Requirement Determination

Requirement Anticipation

Having had a particular business area or having encountered system in an environment similar to the one currently under investigation will influence system analysis study. Requirement anticipation can be mixed blessing on one hand, experience from previous studies can lead to investigation of areas that would otherwise be unnotices by inexperienced analyst. Having a background to know what to ask or which aspect to investigate can be substantial benefit to the organization.

Requirement investigation

This activity is the heart of the system analysis, using variety of tools and skills analyst study the current system and documents its feature for the fixture analysis. Requirement investigation relies on the fact finding techniques and includes method for documenting and describing system feature is following.

- Interview
- Questionare
- Observation
- Record Keeping

Requirement specification

The data produced during the fact finding investigation are analyzed to determine requirement specification, the description features for a new system. This activity interrelated parts :

- Analysis of factual data
- Identification of Essential Requirements
- Selection of Requirement Fulfilment Strategies.

Existing System

This is second activity of the analysis phase. In this activity, the existing system is carefully examined. The assignment of the system is done on the basis of a number of factors like what we are already having and where we want to improve, what has to be done from the scratch etc..

The present MIS system comprises of the information in the form of reports. The marketing personnel daily collects the price but this information is not easily accessible to the producers. Though in our country trader community are well connected. They are well equipped with latest technology like internet, email, fax, telephone etc. traders is playing with information only. The trader community also does not share the information for the benefit of the farmers and only other way our farmer community is highly unorganized and illiterate. Language is also largest barrier in dissemination of such information among the farmers.

The problem with the information which is collected daily is that there is no systematic way of dissemination of information. The directorate of marketing state agricultural marketing board and national horticultural board for horticultural commodities are collecting information and then compiled in the form of weekly, monthly reports. This information is a post mortared data having a very little value and not to percolate to the farmers. The staff for the collection of such information is also not well versed about the techniques of data collection. The information which is being collected covers a very few parameters like maximum, minimum and average price and arrivals. Price alone can not be the only factor for better marketing of a product, other factors are also required to reduce the marketing cost like transportation, packaging, storage, grading etc.

Proposed New System

Project planning is very important aspect of the analysis phase. Details study activities are planned and outline of the remaining system, analysis phase activities are noted. Planning infact is an ongoing process of system development. After having seen the need for such a system, we planned for the Web Market Atlas. The basis planning process includes planning regarding various types of database, which were to be used. Once the planning for the procedure file is done, then we established the relationship among them and an integrated system was modelled.

Integrated system

The online market information is perceived to be national data communication network where by farmers intending to sell a commodity can know the prevailing prices of the commodity in all the markets yards of the country through internet. Farmers undoubtly have a need for information about probable supplies and prices in order to make decisions when planning there production and sales. With the help of information traders help growers to grow with the growth of trade. This reduces their business risk and enables them to operate profitable on lower margins. Thus it brings benefits to producers and consumers. Consumers purchases also can be influenced by market news. They can be encouraged to leave products that are in short supply and by alternatives that are plentiful. This helps in equitable distribution; universal flow pattern and reduction in fluctuation in prices.

The recent development in internet and communication has important applicability in agricultural marketing. Marketing of any produce requires fully equipped and latest market trends. This information is useful to have an access to price prevailing in the market. The markets in turn increase the bargaining power of the producers and it also enhances the efficiencies of the market. The networking concept is applicable to the wide networks of the regulated market.

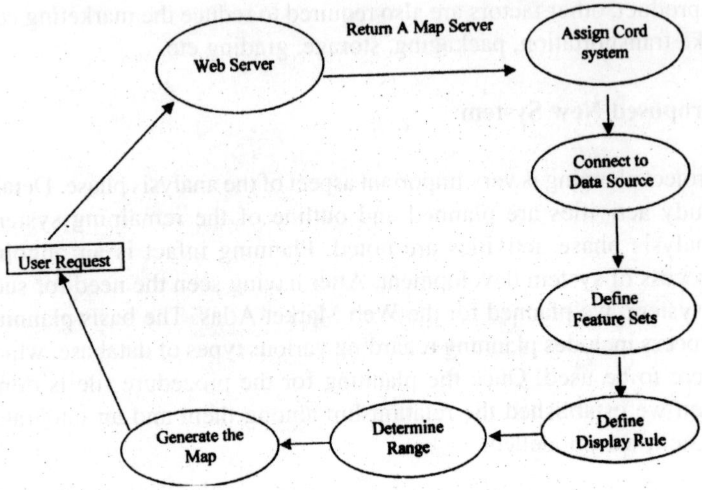

Fig. 44.3 : Generating a Map

The networking would facilitate the flow of information and this information can be easily updated.

The MIS can have the following impact.

- It can facilitate efficient allocation of productive resources.
- It improves the bargaining power of the farmers with the traders.
- Farmers with timely and reliable information and the ability to interpret it can decide to which market they should send their produce to maximize their returns.
- The essence of the good MIS is that it should provide commercially useful information on a timely basis. Information produced on MIS is also useful for the policy makers. This should in the long run improve policy formulation as the functioning of market comes to the better understood.

Market information system is also an important component of Early Warning System for food security as it can assist in identifying areas of possible shortage and can highlight whether prices are below or above normal seasonal trends.

Decisions — After survey of the market areas it was decided to create 28 thematic maps for covering Karnataka and its 27 Districts covering road, categorised by National Highway, State Highway and other roads, Railways, market locations etc.

Future Enhancements

The work on designing the Atlas is in progress . The first map was created for the state of Kerala. The suggestions for the State Government are taken care and necessasry changes are made. Looking into the future developmenets and requirements the same Atlas will be improves by taking following points in consideration:

- Along with shortest distance, path should be traced so as to give a route also between the two APMC's.
- Comparative study of crops arrival, maximum price, minimum price, average price etc. can be done in more ways.
- Retail prices of commodities is also to be linked with each commodity.
- Designing a control panel for modifying static entries is to be made.

SECTION-V

ENVIRONMENTAL CHANGE AND IMPACT

45

The Global Environment

Crossing National Boundaries

Victor Roger Svaage

Introduction

It seems ironic that in an era of globalisation and information technology when the global community is supposed to come together with greater interaction and awareness, we are faced with widening disparities of wealth and the digital divide between countries and between peoples within states. Globalisation and its impact on the national economy has been well documented (Sassen, 1941; Hirst and Thompson, 1996; Dicken, 1998). Similarly, much has been written about the ramifications of globalisation on local and indigenous cultures and their dialectical relationships (Robins, 1997; Hall, 1992). However, comparatively less attention has been devoted to the extent to which globalisation, in all its guises, impinge upon the environment. Unfortunately, whether we deal with the rich or the poor countires or the wealthy and the depraved peoples within societies, the environment at the global, regional or national levels seems to be the ultimate victim.

For the rich countries, the insatiable demands for material goods and conspicuous consumption has threatened the world's natural resources to the point of extinction and exhaustion in some countries in the world. In 1996 for example it was estimated that 25 per cent of the world's approximately 4,630 mammal species and 11 per cent of the 9,675 bird species were at significant risk of extinction (UNEP, 2000 : 4). The demand for timber has left many African countries bereft of trees and vegetation. In addition, many holistic eco-systems are also threatened so much so that the term "bio-diversity" has become a symbol of maintaining ecological equilibrium at global and local levels. The unfortunate driving forces are in rich developed countries seem to be a product of four variables :

1. The capitalist system,
2. Material consumption patterns
3. Urbanization,
4. High energy driven technologies.

In recurrent mantras, we hear that the United States with barely 3 per cent of the world's population consumes something like 40 to 50 per cent of the world's natural resources. Such high consumption patterns are fuelled by material consumer demands, rising expectations, higher living standards, a technologically dependent society and the capitalistic greed to produce more and to make more money out of industrial or service outputs. The United Nations Environment Programme prescribes that if the developing world is to have equitable access to resources for their development agendas, the industrialized countries will have to adopt a ten-fold reduction in resource consumption (UNEP, 2000).

The unfortunate scenario is that the unsustainable demands of capitalism are being pontificated by the developed countries as the panacea for underdevelopment in many developing countries. The collapse of the Soviet Union and the spread of capitalism in Eastern Europe, China, Vietnam and now North Korea might be plus prints for national productivity and increasing levels of per capita income. The down side is that the uneven spread of capitalism in many of these former communist regimes has widened income disparities, created unsustainable methods of production and enhanced individual levels of material expectaions. Capitalism has indeed become the opium of the masses in many socialist societies. The drive to make more money, to consume more and to own more material goods is laudable from a national economic perspective but certainly undesirable in trying to ensure a holistic balance between society and the environments such societies live in. One wonders whether the senseless accumulation of material wealth is only measure for the quality of life ànd human satisfaction.

One of the measures that has been used by academics to look at the uneven consumption patterns lies in the concept, the "ecological footprint". This concept demonstrates that many urban pockets of wealthy people with high consumption patterns is actually being sustained by a vast national, regional and even global hinterland. Hence if you take the living standards of populations like New York, London or Tokyo, it becomes apparent that the inhabitants in such global cities are actually dependent on a vast array of hinterlands around the world. The question obvisously is whether all these materialism and high consuming material goods ad infinitum to legitimize our social status, symbolize our economic clout or endorse ourselves as products of high culture.

It is evident that one of the major contributors to increases in air pollution (carbon dioxide, carbon monoxide and sulpher dioxide) is the ubiquitious presence of motor vehicles especially in urban nodes. For example global emissions of CO_2 have reached a new high of 23,900 million tones in 1996 which is nearly 4 times the 1950 total (UNEP, 2000 : 4). The car might be a means of transport convenience in developed countries but for many nouveau rich societies in developing countries it is the symbol of wealth and affluence. In rapidly developing countries, in the Asia Pacific, the drive for many people to own cars is not only increasing air pollution in urban areas but creating massive traffic jams that are economically unproductive. This situation is difficult to rectify even when we recognize the car as a major culprit of pollution. Why? because in countries where the car is a key industry, many governments actively promote car ownership in order to keep their national car industries productive. Malaysia's national car the Proton is a good example of this sense of national pride and economic bind that the government is in.

The Developing World

The unfortunate situation in the developing world is that while globalisation and information technology is integrating peoples, products and services with greater speed, it is also alienating societies because of the widening disparities of income and wealth arising from these forces. The issues of the developing world have not changed. This is the population that suffers from malnutrition, starvation, disease, illiteracy and unsatisfactory modes of habitation. What are indeed the driving forces that has created this situation, I would suggest three major agents of these problems :

1. Unchecked population growth,
2. Inability of countries to find employment,
3. Ineffective political leaders and governments.

Global Populations Issues

The issue of environmental degradation is tied in with a complex relationship of other issues. The first deals with population growth and the earth's capacity to sustain population. At 6 billion today, the world population is growing now faster than any other time in human history. But its best the rapid population growth that is worrying *per se,* it is the widening gap between rich and poor countries that seems to be a major cause of growing instability. Despite the fact that the global per capita income has passed US$ 5,000 a day, today over 1.3 billion people or 22 per cent of the world's population survive on less than US$1 a day. Another 1.6 billion people or 27 per cent of the world's population live on US$2 a day or US$730 a year. The widening income gap is not only between countries but between residents even in wealthy cities. In Hong Kong for example, 600,000 of its five million residents live on less than US$600 a month which is barely enough to survive there (Mitchell, 2000 : 24). Globalisation is unfortunately widening the gap between the rich and poor. The poor are living in already depressed environments and the poverty and degraded environments create a vicious cycle of unending poverty. The sad fact is that developing countries instead of spending money on issues to alleviate their poverty and plight spend considerable money on defence issues. In 1997, a US study showed that developing countries spent US$232 billion on military expenses or 28 per cent of the world's total military expenditure.

The issue of population growth is no more a local issue. Countries that have severe environmental problems seem to be expanding fastest and this growth is undermining regional ecosystems. While the alarms of the threatened carrying capacity of the earth have subsided since Erhlich's book on the '*Population Bomb*', the issue of unsustainable population growth in developing countries remains a treat to political instability. Given that the cold war has thawed, local and regional problems have not escalated into limited war zones (a la the Korean and Vietnam wars). But there are new regional powers that are willing to fan local conflicts.

In some ways it seems unfortunate that the developed countries have grown tired of the problems in the developing world and hence have shown less concern for them. The attention in the Western world seems to be the preoccupation at various times of the rising threats to their economic and political clout in the global arena with Japan, China and the regional powers of the Middle East. The developed countries seem only interested in issues that involved their own economic interest.

Global Consumerism and Environmental Degradation

Globalisation and its attendant form of global consumerism ('Macdonalisation', 'Cocalisation') is an increasingly entrenched feature within national and local contexts. With the great strides made in the global communications realm (e.g. by now ubiquitous cable televison), the subsuming of global citizens (especially people from the less developed countries) into such a phenomenon through amongst other things advertising can only increase. It has even been pronounced that 'Asians have become full-fledged members of the energy-intensive, mass-consumption, mass-refuse society' (Japan Environmental Council, 2000:7). Clearly as hinted by the quote, such global consumerism is closely aligned with the "throw-away" culture which persists till this very day. This wasteful culture is typified by an astounding proliferation of consumer goods which add very little real utility to one's life and which themselves require much needless packaging. The direct consequence of this rise in global consumerism is the increase in environmental degradation worldwide, particularly in Third World nations.

Many Third World nations in fact bear twice the impacts. Firstly, as the providers of raw materials to fuel this global consumerism and secondly, as part of the very consumers who sustain it. A case in point would be the increased demand for tropical timber around the world—particularly stremming from developed nations. The high demand for tropical timber inevitably draws forest-rich Third World countries into the global trade for timber. However, as Pearce and

Warford (1995 : 281) noted : 'industrial economy import products from tropical forest in quantity that encourage deforestation in the exporting country without building any forms of capital'.

For example, with respect to the intensive exploitation of timber products in East Kalimantan, Indonesia, Emel and Bridge (1995) have noted that 'the overlaying of a highly capitalized, vertically integrated forest industry on the pre-existing smaller-scale system has been disruptive to the forest community and its environment'. The social impacts to the community are both a result of ecological degradation and the increased flows of capital and labour required in such large-scale commercial forest resource extraction. At the broader level, the potential benefits that could be derived from such job creation, deliberate under-valuation of profits and exports and limited domestic processing (Gillis and Repetto, 1998). The case of East Kalimantan highlights two principal challenges facing extractive localities. The first is the difficulty in assuring that an equitable amount of economic rent is charged to (and received) from the various interest groups and the local, provincial, central and international levels.

Secondly, the management of income thus earned has to satisfy both short and long term development needs of the locale. The other words, it would seem that Asia's forests are to be permanently subordinated to the needs of the global market which inevitably reduced them to ecologically monocultures in the process (Rush, 1991).

Besides timber, the demand for other raw materials has also increased at rates which cannot be sustained by current supply. The per capita consumption of resources for the average person has been increasing rapidly throughout the years. This is particularly significant in developing countries. It has been projected that total energy and oil consumption in the newly industrialising countries and other Third World nations could be twice their 1990 levels by 2010 (Emel and Bridge, 1995). Besides the rise of global consumerism, advances made in technology in all its guises have also played an important role in the continuing degradation of the environment.

Technology and the Environment

Underlying the processes of globalisation and the intensification of mass-consumerism is advancement in technology. While it is clear that the IT, digital and mass communications revolutions have helped boost global consumerism, the role of technology in driving globalisation is not limited to this. In many cases, a technologically driven globalisation, coupled with the global mass-consumption, has varying impacts and consequences to the environment.

The clearest example is the increasing reliance on genetically modified food (GMF) such as grains or even meat. As evinced by the development of high yielding or modern varieties of rice since 1996 (Rigg, 1989), the search for resilient agricultural goods is not a new phenomenon. However, the possible extent to which GMF impacts upon our daily life has perhaps only recently received mainstream attention. For one, the safety of such bio-engineered food for human consumption appears unresolved. The recent recall of Taco shells (under the Taco Bell Brand) in the United States and around the world clearly illustrates the lack of institutional control with regard to the sale of such food products. Apparently, the corn contains a gene which makes it more pest-resistant has only been approved to be used as animal feed and not for human consumption. Even in the United States where about two-thirds of processed food contain ingredients made from genetically modified corn, soybeans, potatoes or other crops, labelling and safety-testing of such foods have not been made mandatory (*The Straits Times,* 27th September, 2000).

But beyond the issue of safety, this constant pursuit of the next more sturdy grain meant that existing bio-diversities are often reduced. For example, the current varieties of rice used around the world today are less than 15, a drastic drop from a high of 1,400 varieties grown a few decades ago. Certainly, an increasing globalised rice market and the homogenisation of taste in rice are some of the reasons for this state of affairs. It is thus clear that some of the complex problems brought about by particular globalisations highlighted earlier need to be resolved in a cooperative manner amongst the different concerned actors. At the highest level, global cooperation is imperative. Paradoxically, just as globalisation can bring about accelerated degradation of the environment, a rapidly globalising world may hold the key to solving or at least alleviating them.

What can we do?

Good Governance

We need to ensure that there is good governance at various levels of administration. I would suggest that there are nine indicators for defining good governance. These are (a) participation, (b) strategic vision, (c) the rule of law, (d) transparency, (e) responsiveness, (f) consensus orientation, (g) equity, (h) efficiency and effectiveness, and (i) accountability. Good governance requires most of all good political leadership because without leadership society can be viewed like a fish—it rots from the head down. For example, the late Rajiv Gandhi talked about the image of Calcutta as a dying city of the dead. Yet today I am told the city has been cleared of its garbage and itinerant roadside hawkers and traders have been removed. Calcutta today is a revitalized city, a product of efficient administration and civic consciousness amongst its residents.

In Ahmedabad, the city government was declared bankrupt in 1993. But between 1993 and 1995, the city made 750 million rupees without raising taxes. It raised one billion rupees in bonds and was able to turn the fiscal policies around with good and efficient government management.

Civic consciousness and civic pride

While it is important to have good governance, it is equally important that good governance provides the right environment for civil society to flourish. Essentially, we need to encourage in villages, cities and provinces a bottom up approach and mass participation in environmental issues. We can see how the fruits of good governments and civil societies have changed the environments of cities in India.

In December 1994, the city of Surat was devastated by a lethal plague. One million of its residents ran away from the city. By 1996, the city had become one of the cleanest cities in India because all its citizens got into the act of ensuring the realization of the "Surat First" campaign. This campign galvanised the city's residents to ensure that Surat became a clean city. A similar example can be cited for Colombo, Sri Lanka. Once again the city was able to maintain a clean and environmentally friendly landscape by pro-active citizen participation that broke the bureaucratic stranglehold of the city in the past.

Best practices

Throughout the world we have seen how governments, communities and citizens have been able to turn around adverse environmental problems and create environmental assets. Instead of trying to reinvent the wheel, we should try to exchange ideas, techniques and programmes that have achieved success. We need to look to places which have successfully adopted environmentally friendly products and practices and adapt them to our societies and communities.

For example, when Singapore began its litter free campaign in the mid-1960s, many countries and the mass media were skeptical about its ability to make a Third World City clean and green. But after two decades and with visible results, many cities in Southeast Asia have adopted Singapore's anti-littering campaigns. Indeed last year when I visited Bangkok, the Thai tourist guide warned all tourists that if you were caught littering in Bangkok, you have to pay a hefty fine. So the image of Singapore as a "fine city" (a city of many fines) has now been adopted by other cities in the region. Similarly Singapore's campaign to turn its city into a garden city has also been adopted by other cities in the region—Kuala Lumpur (Malaysia) and Jakarta (Indonesia).

Alternative ways of development

Given that capitalism has become the major system for development around the world one might ask whether this is the most efficient and appropriate method for development in developing countries. If all countries were to follow the capitalistic road to development we would see increasing inefficient uses of raw materials, dwindling natural resources, inefficient uses of energy consumption, mass consumerism and materialism, and increases in air, water and land pollution.

Are there in fact alternative methods of development besides capitalism? Twenty years ago, Schumacher in his book *Small is Beautiful* talked about Buddhist economics and the Ghandian way to development. We need to revisit alternative methods of development that can provide equitable creature comforts, release the creative potential of people and ensure a spiritual and humanly satisfying system of living. I do not have the answers to this mega issue but I think that there are increasing textbooks that are talking about the Asian ways of development and alternative development. Perhaps there is promise in these developmental methods. If we resign ourselves to an American capitalist system we are likely to not only contribute to dwindling natural resources but also environmental degradation.

Reduce consumerism and materialism

One of the major outcomes of capitalism is the rise of consumerism and materialism that is overwhelming not only developed societies but rapidly undermining societies in the developing world. The whole global system of capitalism seems to be hinged on the issue of productivity. In essence, industrial productivity is meant to increase more goods and a variety of goods for the consumer market. Peoples' expectations are increasingly raised by the plethora of consumer goods that flood shops and supermarkets. Consumerism has become the opium of the masses. People feel they are civilized or cultured by various material symbols that they have acquired. This acquisition has spurred a "Throw away mentality" that Alvin Toffler talked about decades ago in his book *Future Shock.* Consumerism is the cause of dwindling natural resources on the one hand and the massive refuse and garbage disposal problems that affect environments on the other. What the world needs is a strong message to de-emphasize materialism and consumerism. The greatest problems that threaten ecosystems around the world is the demands made by materialistic desires. Conservation is not about saving forest reserves and nature areas. Conservation is equally about putting a stop to individual consumer demands. If we recycle and reuse goods, and lower our consumption habits, the sum total of this exercise will have a greater impact on saving nature reserves, natural resources and reducing refuse and garbage.

Energy efficient technology

The recent rapid rise in oil prices that has displayed great discomfort in Europe, the USA and non-oil producing countries clearly demonstrates the fragility of how the world's economic system is governed by one source of energy—petroleum. The price of oil provides mix reactions between countries that are importers as opposed to those that are exporters of petroleum. Yet the gretest beneficiaries of oil prices are essentially the oil companies. And because they reap tremendous profits from oil they have been less than inclined to find alternative sources of energy. We need to wean away the motor vehicle industry from oil dependency. Given that cars are one of the biggest contributors to carbon dioxide and carbon monoxide pollution we need to ensure that this oil dependency by motor vehicles is reduced. Furthermore oil-producing countries generate two to eight times more hazardous waste per capita than the Unites States.

Finding alternative renewable sources of energy can also have major ecological implications. We have seen the ecological impacts resulting from the Aswan Dam in Egypt. We have seen the similar debate that has taken place amongst environmentalists and social scientists about the Three Gorges Dam project in China and we are also passive observers to the controversy of the Bakun Dam project in Malaysia. Once again the quest for capitalist modes of development in Malaysia has overweighed the calls for maintaining ecological sanity in Sarawak. As countries in the developing world latch on to capitalist modes of development, they will be increasingly confronted with the trade-offs between attaining developmental goals ensuring eco-friendly environments.

Environmental NGOs : Tripartite Relationships

Rather than engage in purely environmental groups that crusade agaisnt private companies and government policies, we need to form environmental non-government organizations (NGOs) that represent all three groups : government officials, private sector representatives and environmental groups. This will allow a more effective, constructive and pro-active engagement for the environmental cause. By staying outside of official policies and the private sector economic landscape, environmental groups are wasting a lot of time and energy in non-realistic demands. We need to tap into the private sector because they govern and influence the economic landscape. For example in 1996 overseas development assistance was less than US$ 50 billion but private sector investments totalled about US$ 250 billion (UNEP, 1999 : 2). They can be pressure groups but they may end up as purely voice boxes for political rhetoric. More talk but no effective action. In the globalising world and its complicated ramifications across the board, environmental NGOs need to change their game plans and try to find new ways and alternatives to engage people in a meaningful way to their cause.

In Singapore, the Singapore Environment Council (SEC) represents such an organization. This NGO of about five years old is headed by a private sector tycoon and comprises a committee representing private sector company officials, government officials, academics and green NGO officials. The SEC is funded by both private sector donation and government inputs. Instead of being confrontational, the SEC tries to ensure a balanced approach in dealing with environmental issues.

The Dialogue of Civilizations

When Samuel Huntington presented his article *The Clash of Civilizations* and elaborated on it further in his book of the similar title, he generated a sense of unease around the world. For Huntington, the new world order was not going to be dictated by ideological differences but rather the clash between different broad civilization cultures. For him, notwithstanding the eight broad culture zones in the world, the clash of civilization for the 21st century was to be a conflict between European culture and the Islamic-Confucian societies. Such a dismal scenario, underscored Huntington's fears that western civilization was under threat and that if the West did not wake up to the realization they would lose their current cultural, political, strategic and economic supremacy to the new cultural masters identified as the Islamic-Confucian societies.

Notwithstanding this simplistic deduction, it is clear that Huntington's work has sensitivised many western countries and in particular the Americans to the worry of losing their dominant superpower roles to the new entrants in the global political landscape—Japan, China and possibly India. In a globalised world, there is much that needs to be done in harnessing the cooperation of various cultures to ensure the common good of humankind. As a response to Huntington's thesis, the United Nations has declared the year 2001 to be the year of the "dialogue amongst civilizations". This is an important step to realizing the need that we should minimize conflicts and enhance cultural understanding and cooperation among societies. Wherever we live on earth, it is still the same earth and home that all of us share irrespective of race, creed and culture. Many environmental issues are not location specific and hence demand a broad response. For example in a recent survey of scientist around the world seeking to identify the major environmental problems of the 21st century, the number one concern was "climate change" (UNEP, 1999 : 13)—an environmental issue that will affect all of nature and humanity. We need a collective response to such a global issue. Globalisation and information technology clearly underscores the inter-dependence of societies around the world and that is, certainly endorsed by the fact that we all share the same living environment that makes all life, human and non-human possible. We need to ensure that societies constantly comprehend cultural, linguistic and ethnic differences to ensure a better understanding of one another.

If there is one lesson that we need to learn from Mother Nature, it is that diversity is a strength and not a weakness. The faunistic and floristic diversity that is so evident in the tropics provides the earth not only its air, its climatic stability, its natural resources and pharmaceutical products but also is a source for inspiration and aesthetic relief. Like environmental diversity, we should enhance and maintain the cultural diversity on earth. Each community through a long association with specific environments has developed a repository of knowledge that is hard to replace. The shamans of the Amazon forest tribes are indeed the best examples of walking libraries about the complex tropical forest. When shamans die without the translation of that knowledge to other people, a whole library of knowledge is eradicated. The dwindling of tropical forest and in turn the undermining of many forest dwelling tribes is a poignant reminder of how short-sighted we have become in advancing capitalist modes of economic development.

Education

Whatever stage of life we are in or whatever profession we may hold we remain students throughout our life. The minute we stop learning and become less inquisitive about the world around us we have terminated our life long quest for creative living. In short, education is an important pre-requistite for human life. As an educator myself, I feel strongly that both for short term and long term goals education in various forms is a necessary pre-requisite towards ensuring a harmonious and balanced relationship between human beings and their ecosystems.

For many environmental issues we need to transfer appropriate techniques, knowledge and technologies to help solve environmental issues. This transfer of knowledge and techniques can only be done through informal and formal channels of the educational process. When societies are enlightened by new methods and ways of operating in their environment they will be able to undertake appropriate eco-friendly relationships. As President of the Commonwealth Geographical Bureau, I can safely assure members in developing Commonwealth countries that the transfer of technology and knowledge is one of cornerstone agendas of the Bureau. In 1999, we held two successful workshops on Food Security in Africa and in November 2000, we did two more workshops on Urban Issues in India and Sri Lanka.

In the new millennium, with growing income disparities and the digital divide, we need to be cognizant of the importance of trying to narrow these gaps through various forms and methods of the educational process. It is important to ensure that we do not translate education as purely knowledge, information and awareness of environmental issues. What we need to do and which is lacking in many countries around the world is to ensure that we able to effect *environmental behaviour* of individuals and societies. For example, in many studies in Singapore and some that I was involved in, I have found that both Singaporean students and the general public have a high awareness of environmental problems, both local and global, but what they lack is pro-active environmental behaviour. Few Singaporeans show consciousness in buying environmentally friendly products in retail shops or supermarkets.

Conclusion

At one level, it seems evident that nature has always flourished at three levels : it is a global phenomenon and player whether you take the seas climate, vegetation or soils. Second, besides the global phenomenon of mother nature, there are highly specialized ecosystems and regional niches. If there is one lesson we can learn from nature, it is the importance of diversity (biodiversity) in creating stability and biological innovations. Third, nature is never static. It is the very agent of change and yet it also responds to changes in various ways. This adaptability of nature to constant changes is a lesson we need to learn. In short, Nature has three "D qualities" we must keep in mind : it is dynamic, diverse and distinctive.

These three lessons from Mother Nature need to be understood before making decisions on the modification and destruction of our environments. We often take an anthropocentric and egoistical vantage point about our relations with nature. We see ourselves as the dominant species with the highest intellect and having the closest relationship with God(s). But we need to be reminded that we share this place called Earth with thousands of other species and billions of other organisms. If we oust the other organisms for ourselves and our engineered landscapes, God above knows when we as a species will be facing a major backlash for our reckless behaviour. Time will be a determinant. We have seen around the world how reckless developments have created major problems for human communities. The massive Yangtze floods in 1999 and Mekong floods this year (September/October 2000) are poignant examples of how deforestation is creating ecological instability.

Globalisation and global consumerism has led to the rapid depletion of natural resources. As argued in this paper, globalisation signals not only an increasingly interdependent world but it also has uneven impacts on the environment. To be sure, the full extent of the ramifications of a increasingly 'modified' environment is yet to be known. What is clear though is that the benefits derived from all forms of exploitation are uneven in their distribution. To the extent that environmental problems are transboundary, there is an urgent need for the international community to address these problems within a more coherent framework. If anything, the complexities of environmental issues in an increasing globalised world requires greater resolve and determination from all concerned actors more than ever before.

In the 1980s we were governed by Rene Dubos' dictum to "think globally and act locally". Today, with the full swing of globalisation, we need to think the other way around as well. We need to think locally about all the ecosystems that are under threat from globalisation and how we need to save them and preserve biodiversity. In doing so, we need to act globally and ensure international and regional steps be taken to halt the global impacts on local and provincial biosystems.

In this on going endeavour, I like to close as an educationalist with a quote from the great Chinese sage Confucius about the value of education :

If you want to plan for one year, plant rice.
If you want to plan for ten years, plant trees.
But if you want to plan for 100 years, educate the young.

REFERENCES

Mitchell, Mark (2000) 'Poor Show' *Far Eastern Economic Review,* October 26, p. 24.

Robbins, K. (1997) 'What is the World is Going On' in Paul du Gay (ed.) *Production of Culture/Cultures of Production,* London Sage in association with the Open University.

Rigg, K. (1989) 'The Green Revolution and Equity : Who Adopts the New Rice Varieties and Why?' in *Geography,* Vol. 74, Part 2, No. 323, pp. 144-50.

Rush, J. (1991) *The Last Tree-Reclaiming the Environment in Tropical Asia,* New York : The Asia Society.

Sassen, S. (1991) *The Global City : New York, London, Tokyo,* Princeton University Press.

Savage, V.R. (1999) "No Smoke Without Fire Dealing with the Southeast Asian Haze" in Awang, A., Salim, M. and Halldane, J.F., Rahman, A. (eds.) *Applied Information in Environmental and Urban Management,* Kuala Lumpur : Institute Sultan Iskandar, Urban Habitat and Highrise Monograph SEACEUM 5.

The Straits Times, 27th September 2000, Singapore : Singapore Press Holdings.

UNEP (1999) *Overview Geo-2000 Global Environmental Outlook,* Kenya : United Nations Environment Programme.

46

Environmental Degradation and Land Use Changes in Saline Water Eco-System in Southern Bangladesh

Abu Muhammad Shajaat Ali

Like most Third World countries, environmental degradation is a major impediment to economic growth in Bangladesh. Numerous studies dealing with the issues of environmental management in Bangladesh indicate that the country experienced loss of soil quality and crop yield, deforestation, soil erosion, depletion of water resources, extinction of fish species and a declining livestock population over the past decades (Alam, 1995; Elahi, 1992; Fallon, 1990; Haque and Hossain, 1988; Ipe, 1995; Jansen, 1986; Mahtab and Karim, 1992; Mofizuddin, 1981; Saleh and Nishat, 1988; UNDP, 1989; Winterbottom, 1990). Much of this environmental degradation scenario is biased to the rural populace and affects the small holder subsistence farmers who possess diverse but deep and intimate knowledge in local ecosystem management problems and adopt various land use and farming strategies to sustain agricultural production. What is missing in the existing studies on Bangladesh environment, but exceptionally important for future national economic development planning, is an examination of the diverse nature and causes of environmental degradation in microlevel ecosystems and the diversity in farmers' knowledge and strategies they adopt to cope with the environmental degradation and sustain agricultural production. The present study aims to examine the nature and causes of environmental degradation in a village situated in the saline water ecosystem in south-western Bangladesh. It also examines the village farmers' knowledge in environmental degradation and their strategies to cope with those constraints and land use changes that emerged from those strategies.

Environmental Degradation, Farmers' Knowledge and Sustainability of Agriculture: A Conceptual Framework

Environmental degradation is a human induced negative consequence of complex interrelationships between people, nature, society and technology (Fig. 46.1). At the national scale, environmental degradation results from growing population pressure, poor socio-economic conditions, regional geopolitics, human induced global and regional climatic changes, and technological change (for details, see Sadik 1988; Lipton, 1977; Lonergan, 1993; Mellor, 1988; Myers, 1988; Goldemberg, 1993; UNCED, 1992; Hulme and Kelly, 1993; Commoner, 1988; Douglas, 1994; Tematio and Olson, 1997). At the micro-level, environmental degradation in rural ecosystems results from factors that are both endogenous and exogenously induced and concerns farmers. Endogenously, growing population pressure, landlessness and poverty force small holder farmers to over-cultivate, over-graze and clear forests to increase food, fodder, fuel and timber production and causes soil erosion, deforestation, depletion of ground water resources, and loss of crop yield (Sadik, 1988; Lipton, 1977; Lonergan, 1993; Mellor, 1988; Myers, 1988). At the farmstead level, environmental degradation varies inversely with farm size (Lipton, 1977; Lonergan, 1993). Exogenously, factors of international (geopolitical) and national dimension proportionally trickle down into the rural ecosystem and cause long term and irreparable damage to the environment. Such factors may include international and national strategies of resource use, technological innovations with superior production capabilities. The gross environmental degradation (GED) is a function of the combined effects of' both endogenous and exogenous factors: growing population demand by socio-economic condition (farm size), increases the intensity of resource exploitation *via* using high volume of exogenously induced technology inputs under the influence of exogenous strategies, and causes environmental degradation.

Sustainability is a technical concept that aims to recover the environment from degradation, increase agricultural output, achieve economic growth, and preserve the environmental quality to guarantee equally suitable natural resource base for the future population (Beckerman, 1994, Benbrook, 1991; Azar, Holmberg, Lindgren,

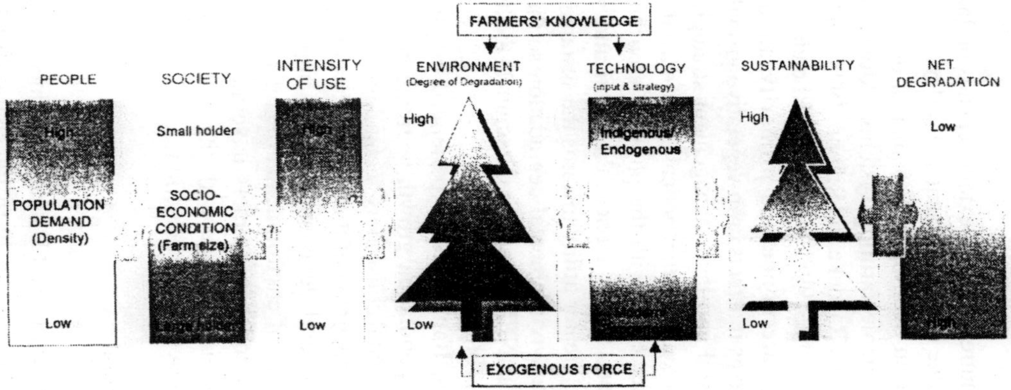

Gross Environmental Degradation (GED) = Pop. Demand/farm size x Intensity of Res. Use x Modern tech. inputs[1] x exogenous forces[1]

Sustainability (S) = Pop. Demand/farm size x intensity of Res. Use x Indigenous tech. Inputs[2] x exogenous forces[2].

Net Environmental Degradation (NED) = (GED – S)

1 Both endogenous and exogenous forces including modern technology inputs cause irreparable damage to the micro-ecosystem. Indicators of degradation are loss of yield, soil erosion, deforestation, loss of soil quality, increased flood, drought and sedimentation.

2 Indigenous technology and strategies, and minimal influence from exogenous forces may minimize the human induced damages and help the ecosystem to recover from the damage. Indicators of sustainability includes growth in yield, land and labor productivity, increased cropping intensity, maintenance of soil fertility, and economic growth as indicated by increase in per capita income.

Fig. 46.1 : Environmental degradation, farmers' knowledge and sustainabillity of agriculture in a micro-ecosystem : a conceptual

1996). Sustainable rural systems show gradual increase or stability in land, labour and marginal productivity in agriculture, growth in farm income, socio-economic development, and gradual improvement in soil quality and fertility.

Sustainability in agriculture to a great extent depends on farmers who possess a deep knowledge in environment and soil management and plays an important role in averting environmental degradation and developing more sustainable farming systems (Altieri, 1987; Alvard, 1993; Eyzaguirre, 1992; Raedeke and Rikoon, 1998; Richards, 1985; Rhoades, 1992, 1994; Sinha, 1998; Talawar and Rhoades, 1998; Thurston, 1990). Farmers are concerned primarily about the topsoil-moisture relationship as it affects crop vegetative growth and yield. Their knowledge in soil and land management strategies are acquired through generation long *in situ* practical experience that are reflective of their close interaction with the physical environment (Krogh and Paarup Laursen, 1997). Hence, their soil and water management strategies have greater utility towards the sustainability of agriculture as they reduce environmental degradation so that the net environmental degradation (NED) would be minimum in the highly sustainable farming systems (Fig. 46.1).

Over the past 50 years, Bangladesh experienced major land degradation owing to rapid population growth, extreme land stress and poverty, severe environmental constraints, technological changes in agriculture, and due to the diversion of the Ganges River water by India. For sustainable agricultural development, it is important to examine the degree and nature of environmental degradation at the micro-level rural ecosystem and how farmers have adjusted their farming and land use strategies to cope with the environmental degradation and sustained agricultural production.

Study Area, Data and Method

The study was conducted in the village of Damarpota, a medium sized village (388 ha) situated in the floodplain of the Betravati *(Betna)* River (a distributary of the Ganges River), 4 km east of Satkhira town in south-western Bangladesh (Fig 46.2). The village has a population of 1,400 living in 139 family households. There are 19 large holders (>3 ha each), 21 medium holders (1-3 ha), 44 small holders (0.2-1 ha), and 55 landless (< 0.2 ha) farm households in the village.

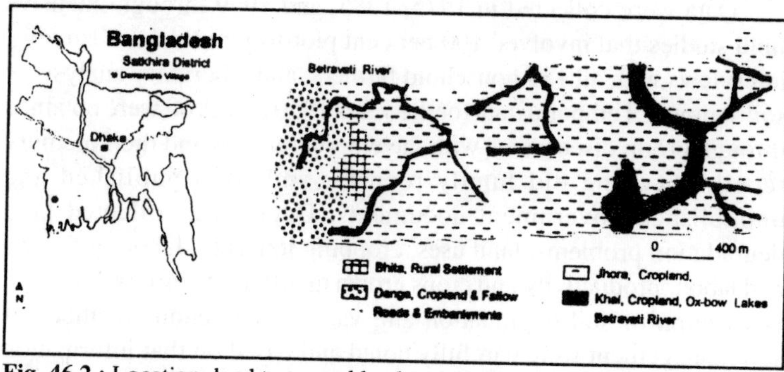

Fig. 46.2 : Location, land types and land uses in the village Damarpota, 1999.

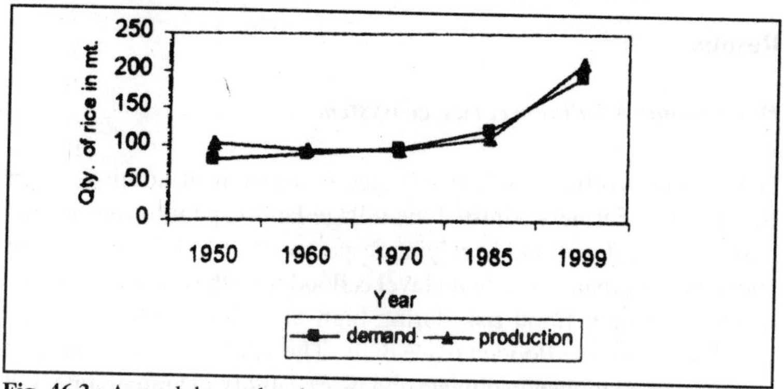

Fig. 46.3 : Annual demand and production of rice in Damarpota : 1950-1999.

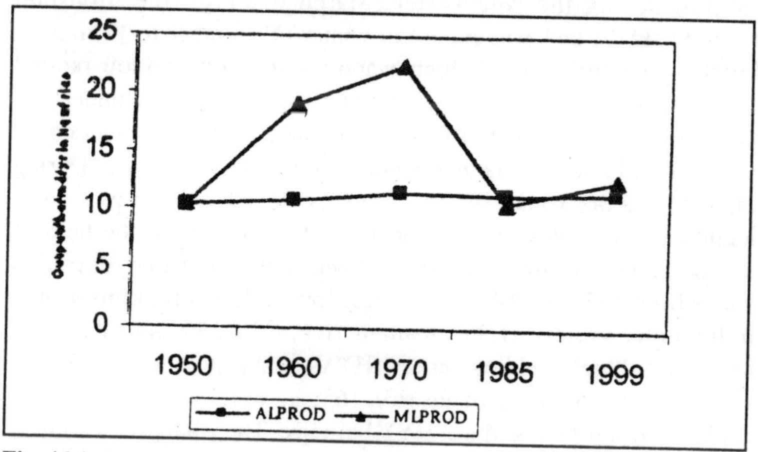

Fig. 46.4 : Return to labour inputs in rice farming in Damarpota : 1950-1999.

Data were collected in 1975, 1985, and 1999 through intensive field studies that involved 100 per cent plot-to-plot land use surveys, interviews with all 139 household farmers, and laboratory analysis of soil samples. Temporal data for the years 1950 and 1960 were obtained through group discussions with the elderly farmers and their memory recall, checking old family records, and from published and unpublished government documents. Data were collected land degradation problems, land uses, cropping intensity, labour use, land and labour productivity and crops grown in different seasons. Farmers' description of soil degradation and various indigenous methods of soil management were carefully noted and based on that information a land degradation map was prepared.

Results

Damarpota: A saline wet rice ecosystem

Damarpota typifies a saline wet rice ecosystem of south-western Bangladesh. Farmers adjusted the village landscape into four levels, *bhita, danga, jhora,* and *khal* lying respectively at 4, 3, 2, and 1 metre above the riverbank. The *bhita* level is flood free; the *danga* and *jhora* levels have high flood risk during high monsoon rainfall; and the *khal* level is tidally flooded twice daily. The village farming depends on the tropical monsoon climate and its variability in temperature and rainfall during the Nor'wester (March to May), the Monsoon (June-October) and post-monsoon winter (November to February). High temperature (+39°C), high evapo-transpiration and low rainfall (<50cm) during the Nor'wester season cause drought that hinders the *kharif season* cultivation of Aus rice and jute. High monsoon rainfall (> 140 cm) facilitates the *haimantic* cropping of Aman rice. During the cold and dry winter season, farmers cultivate *rabi* crops such as legumes, vegetables, wheat, and HYV boro rice with the help of low-lift pump irrigation. Labour requirement (per ha) in rice cultivation varies from 135 man-days for aus rice, 150 man-days for transplanted indigenous variety (LT) of aman rice, 220-250 man-days for transplanted high yielding variety (HYV) aman and boro rice. Per ha yields vary from 800 kg of aus rice, 1650 kg of LT aman, 2500 kg of HYV aman and 3500 kg/ha of HYV boro rice. The village environment has degraded owing to human induced endogenous and exogenous factors.

Endogenous sources of environmental degradation

People and Society

Despite high soil salinity and low agricultural intensity, Damarpota population grew 2.5 times (from 483 to 1206) between 1950 and 1999 when the physiologic density grew from 124 people km^{-2} to 361 people km^{-2}. Growing landlessness and major resource polarization followed population growth as the per capita availability of arable land declined from 0.39 ha to 0.11 ha and the amount of farmland under small holder category declined from 35 per cent to 13 per cent and that under large holder category increased from 13 per cent to 41 per cent during the period in question (Table 46.1). Growing population pressure and landlessness forced farmers to over-cultivate their land causing soil erosion and declining crop yield.

Exogenous sources of environmental degradation

Technological change in agriculture

The Green Revolution was launched in Damarpota area in the mid-1970s when high yielding varieties (HYV) aman and boro rice seeds, Nitrogen-Phosphorus-Potassium (N-P-K) based chemical fertilizers, pesticides and mechanized low-lift pump irrigation were introduced. Low lift pump irrigation allowed expansion of HYV rice cultivation during the dry seasons and uses of chemical fertilizers and pesticides increased the total rice production in the village (Ali, 1987). However, frequent cultivation of HYV rice, intensive use of chemicals, and irrigation water have seriously damaged the aquatic habitat, degraded soil quality, and lowered animal fodder production.

Shrimp farming

In the early 1980s, the village's saline ecosystem attracted the outsider shrimp farmers who leased the village farmland and transformed the rice fields into shrimp farms, locally called *gheres,* by pumping saline water from the Betravati River. To maintain high level of water salinity in order to facilitate rapid shrimp growth, additional salts were also added to the shrimp farms. Continuous shrimp farming for over 10 years increased the soil salinity to an alarming level as thin salt layers are often seen on the topsoil.

Geopolitics

The summer-time diversion of the Ganges river water by India and massive deforestation in the Himalayan Nepal are the exogenous geopolitical factors that also affect the ecosystem of the study area. Decreased discharge of the Ganges river decreases the seasonal discharge of its distributary, the Betravati river, and causes rapid depletion of the underground water resource hindering irrigated HYV boro rice cultivation in Damarpota. It also increases the soil salinity in the village as the saline Betravati river water tidally floods the village land.

Rapid deforestation and massive soil erosion in the Himalayan Nepal increases the sediment load of the Ganges River System, which in turn causes rapid sedimentation of the Betravati riverbed. Regular tidal flooding with silt-rich river water causes sedimentation of ox-bow lakes and khals that are otherwise used as water reservoir for dry season cultivation and shrimp farming.

Climatic changes

The village experienced major climatic shifts in recent years, which may also be tied to deforestation in the Himalayan mountains and the Ganges water diversion by India. In recent years, Damarpota and Satkhira districts experienced record high summer and record low winter temperature, high evapo-transpiration and declining pre- and post-monsoon rainfall to cause regular drought, and indicate desertification in progress.

Farmers' knowledge in environmental degradation

Farmers are aware of on-going soil degradation in Damarpota. They viewed that the current state of soil degradation is human induced and due to existing inappropriate soil management practices. They see much of soil degradation in the village today is primarily a result of *over-cultivation, over-grazing and deforestation* following rapid population growth and landlessness which forced them to cultivate more frequently and over-graze the fallow pasture lands. Growing need for firewood and building materials led to deforestation of village social forestry by 56 per cent over the period.

Farmers reported minor *soil erosion* due to over-cultivation, over-grazing, field raising, construction of field dykes and inappropriate tillage using power tillers followed by heavy rainfall. Soil erosion increased soil stoniness in about 3 per cent of soils in *danga* land. Severe erosion is also observed along the riverbanks. About 24 per cent of the village soils experienced erosion and loss of fertility. Most farmers (92 per cent) perceived that increased *salinity* owing to shrimp farming is the most important form of soil degradation affecting 80 per cent of the village soils (Table 46.1). Low-lift pump irrigation of HYV born rice fields also increased soil salinity.

Farmers perceived the occurrence of *drought, desertification* (locally called *khora) and depletion* of *ground water resource* due to lack of summer rainfall when underground water table drops by >30m, the village ponds, tube wells and irrigation pumps dry up. They are aware of recent *decline in soil fertility* as indicated by sharp declining of HYV rice yields by over 50 per cent in recent years, which they attribute to soil degradation due to increased soil salinity, soil erosion, drought, formation of chemical, salt, and sediment layers on top soils, and over-cultivation. Farmers also reported that increased cultivation of dwarf HYV rice crops, and expansion of shrimp farms on rice fields rapidly *declined fodder supply* that fostered decline in cattle population from 366 in 1985 to 155 in 1999.

Farmers' knowledge in soil management and ecosystem sustainability

For many generations, the farmers of Damarpota had developed various indigenous soil management strategies to cope with tidal flooding, drought, soil fertility loss and salinity, and maintained a highly sustainable farming practice. To cope with the flood and drought, and manage the surface water resource, they construct flood control embankments *(bunds)*, dig ponds and *khals* as reservoirs, build

Table 46.1: Population pressure and environmental degradation in Damarpota: 1950-1999

Year	*Population*	*Per capita Arable land ha*	*% of land flooded*	*% of land drought affected*	*% of land affected by salinity*	*% of land exposed to soil erosion*	*% of land experienced sedimentation*
1950	483	0.39	45.3	24.4	74.1	1.0	12.0
1960	617	0.30	36.7	23.1	71.0	2.0	16.0
1970	692	0.24	23.7	28.0	66.5	2.0	23.0
1985	887	0.17	23.9	54.0	55.4	2.9	29.0
1999	1206	0.11	23.0	100.0	79.6	4.5	36.0

Source: Authors field works in 1975, 1985, 1999.

Table 46.2 : Land use changes and sustainability in agriculture in Damarpota: 1950-1999.

Year	*% settlement*	*% cropland*	*Cropping intensity %*	*Cropping intensity by farm size %*			*Total Output kg/ha/yr*	*Average Labour output k /md/ha*	*Marginal labour output k /md/ha*	*Food demand mt/yr*	*Total Food Production mt/yr*
				Large	*Medium*	*Small*					
1950	0.70	90.50	96.00	93.00	98.00	97.40	1,338	10.45	10.45	81.00	102.00
1960	1.17	90.00	96.70	91.00	99.00	99.40	1,472	10.91	19.10	89.00	94.00
1970	2.15	89.00	102.10	100.00	102.00	105.20	1,651	11.54	22.40	96.00	93.00
1985	2.81	88.35	111.00	111.40	119.00	123.00	1,991	11.54	10.63	121.00	109.00
1999	3.82	88.00	158.00	139.00	148.00	187.00	2,201	11.78	12.01	192.00	210.00

and maintain water retention walls around each field *(ails),* and adjust field surface slopes to achieve its uniformity. Conservation tillage and soil-moisture monitoring and cultivation of more flood and drought resistant cultivars are commonly practiced here.

The three widely used methods of soil fertility management on rice fields are crop rotation, conservation tillage and cattle grazing on crop residue. Farmers cultivate the aman rice during the *haimantic* (monsoon rainy) season, legumes (beans), oilseeds, and tobacco during the *rabi* (winter dry) season, and leave the land fallow during the *kharif* (summer) to regain soil fertility. They believe that the cultivation of legumes increases soil fertility (scientifically it would mean nitrogen fixation), oilseeds would reduce soil salinity, and that of tobacco would control pests. Farmers leave crop residues on the field and allow cattle and other animals to graze on it to derive multiple sustainable benefits. Grazing on crop residue saves the cost of fodder. The grazing cattle would add manure directly to the field and save the cost of purchased chemical fertilizers. Crop residues and roots remaining in the field would retain soil moisture and save the cost irrigation, while the decomposed plant residue would add organic fertilizer to the field.

One important task of farming in this saline water ecosystem is coping with increasing soil salinity from tidal flooding. Government constructed embankments has reduced the impacts of tidal flooding but naturally occurring salinity and acidity inherent to this soil poses major threat to agriculture. Farmers adopted two strategies to reduce these constraints. For centuries, they applied kitchen ashes to the rice field, which they believed would reduce soil salinity. They also collect and burn clam, snail and crab shells to get lime and apply that to rice field to reduce and balance soil salinity and acidity (*kosh* or *khar*).

Changing land uses and sustainability in agriculture

Despite major degradation in its environment, the village land use patterns and farming systems have changed over the past 50 years in question. As population grew by 2.5 times over the period, area under settlement expanded by five times consuming 9.4 ha of cropland (Table 46.2). Population growth was followed by agricultural intensification despite growing environmental degradation over time. In 1950, 96 per cent of the village farmland was annually cropped to transplanted Aman rice during the *haimantic* season. With the advent of the green revolution technological changes, cropping frequency has increased to 158 per cent in 1999 due to the expansion of irrigated rice cultivation in the rabi season. Frequency of cultivation varied significantly among farmers. Small holders cultivate more intensive than the large and medium holders.

Cropping patterns also changed over time. In the 1950s and 1960s, traditional local transplanted aman rice occupied 98 per cent of farmland. In the 1970s, about 34 per cent of cropland went under HYV aman rice and only 15 per cent was allocated to HYV boro rice during the rabi season. In the 1980 and 1999, local aman rice was completely replaced by HYV aman rice and the cultivation of HYV boro rice expanded to irrigated lands. Low-lift pump irrigation increased the soil salinity and that hindered HYV boro rice cultivation.

Prior to the early 1970s, traditional ox-plow technology and local transplanted aman rice characterized the village agriculture. Since the mid-1970s, green revolutionary technology inputs such as low-lift pumps, chemical fertilizers, spray machine, pesticides, and power tillers became available to the village farmers. Large and medium holders have better access to modern inputs while small holders still rely most on traditional technology to cultivate modern HYV rice.

Both intensity of labour and total output/ha/yr have increased during the period in question (Table 46.2). Until the 1970s, total output/ha/yr (TP), average output/labour input (AP) and marginal output/labour input (MP) had increased (although TP increased only by 313 kg/ha), primarily as a result of flood and salinity control by the construction of Betravati embankments. In the 1980s, increases in total output/ha has been attributed to double cropping with HYV rice, use of chemical fertilizers and low-lift pump irrigation. The results were that total output/ha/yr increased by 300kg/ha/yr. Average output/labour input decreased only slightly, by 0.16 kg/ha, while the marginal output/labour input declined by 50.0 per cent (Fig. 46.4). This indicates that the technological change or the "green revolution" has increased the total output/ha/yr, but at the cost of declining average and marginal output per labour input. In the 1990s, the declining yield due to growing environmental degradation resulted in further decline in labour productivity although the uses power tillers reduced the labour use and saved the ecosystem from agricultural involution in its worst form. As expected, labour input was inversely related with farm size: small holders use the highest intensity of labour/ha, followed by medium and large holders.

During the period in question, food demand has doubled. Food production has increased but not at the same rate. In the 1950s, Damarpota was a rice deficit village owing to uncontrollable salinity and tidal flooding. In 1970, there was small deficit in food production; the demand production gap increased because most of the village land were transformed into shrimp farms. In 1999, there was major surplus of rice production because the farmland was returned to HYV rice farming (Table 46.2). Although the village is self-sufficient in food production, small holders still suffer from food shortages while the large farmers produce a big surplus for the market. Average farm income in *Damarpota* is low. However, over the past decade, average household income of all farmers increased due to increasing amount of off-farm employment in and outside the village.

Conclusion

The study has taken much interest in exploring the nature and sources of environmental degradation in a saline water ecosystem in south-western Bangladesh. Over the past half of the 20th century, the village has experienced major degradation of its environment that affected the farming systems and farm production. Farmers possessed deep and intimate knowledge in the environment and adopted various indigenous strategies to cope with the environmental constraints and still maintain and increase agricultural production for the growing population. The village agriculture shows the sign of sustainability although the environment suffers from degradation and the economy remains under poverty that biased towards the small holder farmers and landless farmers. Opportunities for off-farm income may be the optimum solution to boost the economic condition of the farmers.

47

Growth Responses of Some Ornamental Plants to Simulated Acid Rain

G. Prakash, A. Kumar, S. Mittal, S. Sharma and *S. Poonia*

Introduction

The transfer of harmful amounts of natural and synthetic materials into the air due to man's activity has now a dominant role in our daily life in terms of air pollution. Contrarily nature has given us a life-sustaining environment. Different industries and fossil oil driven vehicles release various noxious gases such as sulphur and nitric oxides (SO_X and NO_X) which on interacting with water vapour in air form acids (sulphuric and nitric) that fall on earth in the form of acid rain. Acid rain has a number of adverse implications, threatening soil fertility, vegetation, water bodies, habitats and fauna. In addition it also corrodes buildings, statues, bridges, monuments, railings, etc. Acidification of soil and lakes due to acid rain allows toxic metals such as aluminium and mercury to be leached from toils and sediments.

Flowers as object of beauty, which bring men in proximity of nature and play a vital role in the life of human beings. Floriculture has become one of the potential money-spinners for developing countries like India. Therefore, the present study was undertaken to observe the effects of acid rain (simulated) on eight ornamental plants belonging to four different families such as Apocynaceae, Asteraceae, Caryophyllaceae and Polygonaceae.

Materials and Methods

For the present experiments, eight different known and available ornamental plants were selected and included—*Catharanthus roseus*

L. (Apocynaceae), *Calendula officinalis* L., *Chrysanthemum indicum* L., *Dahlia pinnata* 'Cav', *Tagetes erecta L.* (Asteraceae), *Dianthus barbatus* L., *Dianthus caryophyllus* L. (Caryophyllaceae) and *Phlox drummondi* L. (Polygonaceae). Except the first one (perennial), rest are annuals and seasonal plants. They are commonly grown in nurseries and gardens.

Acid water solutions of different pH, 5.6, 4.5, 3.5 and 2.5 were prepared by using a mixture of sulphuric and nitric acids in the ratio of 7:3 (V/V) in distilled water (Lee *et al.*, 1981). The pH values were confirmed with the help of an electronic pH meter with a glass combination electrode. The solution of pH 5.6 was taken as control.

The seeds of different selected plant species were pre-soaked in different concentrations of acid water for 6h and then grown in garden soil filled in earthen pots. The saplings on attaining two leaves stage were given regular weekly treatment of simulated acid rain (s.a.r.) as spray till their full wetting.

The effects of s.a.r. with different pH were analysed for the following growth parameters taken at 20d intervals of plant age, like length of root and shoot, fresh weight of root and shoot, dry weight of root and shoot, leaf number per plant, and individual leaf area.

From the data analyzed for the above attributes following attributes were further derived:

$$\text{Shoot Weight Ratio (SWR)} = \frac{\text{Shoot dry wt.}}{\text{Plant dry wt. (phytomass)}}$$

$$\text{Root Weight Ratio (RWR)} = \frac{\text{Root dry wt.}}{\text{Plant dry wt. (phytomass)}}$$

$$\text{Shoot Root Ratio (SRR)} = \frac{\text{Shoot dry wt.}}{\text{Root dry wt.}}$$

$$\text{Net Primary Productivity (NPP)} = \frac{\text{Plant dry wt. (phytomass)}}{\text{Plant age}}$$

$$\text{Per cent Moisture Content (\%MC, fresh weight based)} = \frac{\text{Fresh wt.} - \text{Dry wt.} \times 100}{\text{Fresh wt.}}$$

$$\text{Per cent Moisture Content (\%MC, dry weight based)} = \frac{\text{Fresh wt.} - \text{Dry wt.} \times 100}{\text{Dry wt.}}$$

In addition to above growth related attributes, following biochemical attributes such as leaf extract pH, chlorophyll (Aron,1949), ascorbic acid (Keller and Schwager,1977) and sulphur content (Patterson, 1958) were also analyzed.

Results and Discussions

The different ornamental plants exposed to different concentrations of acid rain showed clear-cut responses for various attributes studied for the impact of s.a.r. The responses were both positive and negative in the sense of increasing or decreasing growth attributes. It was observed that the higher was the concentration of the pollutant, i.e. lower the pH, more detrimental were the effects.

In the experiment, growth attributes associated with root and shoot were studied in all plants. The root system was found to be more prone to s.a.r. than the shoot system as noticed from the reduction in size and weight of the roots. The root length of the plants decreased following s.a.r. treatment in all taxa. The reduction was observed maximum at pH 2.5 and minimum at pH 4.5. The per cent decrease recorded in root length at pH 2.5 in 80d old plants were 35.6, 46.7, 61.7, 30.9, 51.2, 56.4, 25.0 and 54.2 in *Catharanthus roseus, Calendula officinalis, Chrysanthemum indicum, Dahlia pinnata, Tagetes erecta, Dianthus barbatus, Dianthus caryophyllus* and *Phlox drummondi* respectively. Like root length, shoot length also got decreased following s.a.r. treatment in all taxa: The per cent decrease recorded in shoot length at pH 2.5 in 80d old plants were 33.4, 31.1, 39.5, 32.9, 29.0, 44.2, 42.4 and 46.4 in *Catharanthus roseus, Calendula officinalis, Chrysanthemum indicum, Dahlia pinnata, Tagetes erecta, Dianthus barbatus, Dianthus caryophyllus* and *Phlox drummondi* respectively. The impact was acidity based, i.e. lower the pH, more was the reductive effect. Singh and Rao (1979), Kohut *et al.* (1987), Khan and Khan (1991), Gupta et al. (1993) and Verma (1999) have also reported decrease in root and shoot length due to acid rain treatment (Table 13.1 a & b).

Reductions were also observed both in root and shoot weight (fresh/dry both) of all presently studied ornamental plants. Maximum reduction was observed in 80d old plants at pH 2.5 and minimum at pH 4.5. Dry weight fractions of root and shoot also recorded a decrease.

Dianthus barbatus was the most sensitive plant followed by *Tagetes erecta, Chrysanthemum indicum, Calendula offcinalis, Dianthus caryophyllus, Phlox drummondi, Dahlia pinnata* and *Catharanthus roseus* for root. Phlox *drummondi* was the most sensitive plant followed by *Dianthus barbatus, Chrysanthemum indicum, Catharanthus roseus, Calendula officinalis, Dianthus caryophyllus, Tagetes erecta* and *Dahlia pinnata* for shoot. This is attributed to slow translocation of metabolites in the root, photosynthetic activity being depressed by the acidity of s.a.r. Reductions in shoot and root dry weights were initially lesser, while at the final harvest, prolonged acid exposure projected marked decline in dry matter. These findings were similar to those of Kumar (1997) and Verma (1999). The explanation given by Noyes (1980) and Kojiol (1984) suggests that under s.a.r. stress, phloem loading system becomes a major limiting factor to control translocation rate.

The effects of s.a.r. on leaf area and leaf no. of different ornamental plants were also investigated. In treated plants with s.a.r., the green area got reduced because of their leaf size and injury (Fig 47.1 and 47.2). This is closely related with the reduced dry weights of shoot and root. Many workers like Takemoto *et al.* (1988) in *Capsicum annuum,* Linskens *et al.* (1988) in *Phaseolus vulgaris;* Adaros *et al.* (1988) in *Vicia faba* and Jacobson *et al.* (1985) in radish also reported reduction in leaf area due to depression in development of the leaves as a result of acid rain pollution. Reduced size of leaves in the studied ornamental plants is regarded as an adaptive measure developed by

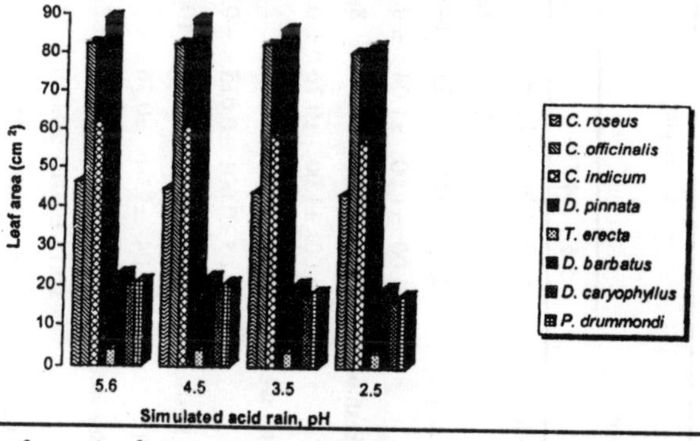

Fig. 47.1: Leaf area (cm^2) in response to different pH of s.a.r. in different plant species.

Table 47.1(a) : Growth responses in terms of length, fresh and dry weight of root and shoot and per cent moisture content, both fresh and dry weight based in 80d old plants.

Attribute	*C. raseus L.*				*C. officialis L.*				*C. indicum L.*				*D. pinnata 'Cav'*			
	(pH of acid rain)															
	5.6	4.5	3.5	2.5	5.6	4.5	3.5	2.5	5.6	4.3	3.5	2.5	5.6	4.3	3.5	2.3
Root length	15.00	13.66	11.66	9.66	10.33	8.50	7.00	5.50	12.66	10.00	7.00	4.66	11.33	10.00	8.33	7.83
	±1.00	±1.04	±0.28	±1.04	±0.76	±1.00	±0.50	±1.00	±0.28	±0.50	±0.50	±0.28	±1.52	±1.00	±1.04	±1.52
Shoot length	22.00	19.00	16.33	14.66	17.66	13.83	12.66	12.16	17.83	15.50	13.00	11.00	14.66	13.33	12.00	9.83
	±1.00	±1.00	±1.04	± 0.28	±0.76	±0.76	±0.76	±1.04	±0.28	±0.50	±0.50	±1.00	±0.57	±1.52	±1.73	±1.52
Root fresh weight	12.00	11.00	10.33	8.66	8.33	7.66	6.83	5.00	9.50	7.83	6.33	4.83	11.83	11.33	9.00	8.83
	±1.00	±1.00	±0.76	± 0.28	±0.76	±0.28	±0.76	±0.50	±0.50	±0.28	±0.57	±0.37	±0.76	±1.52	±1.00	±2.08
Root dry weight	4.03	3.65	3.43	2.87	2.75	2.52	2.24	1.63	3.14	2.62	2.09	1.60	3.92	3.75	3.00	2.77
	±0.05	±0.001	±0.015	± 0.10	±0.015	±0.015	±0.015	±0.015	±0.02	±0.015	±0.01	±0.015	±0.020	±0.015	±1.00	±0.15
Shoot fresh	15.33	14.33	12.66	10.33	18.66	15.66	13.33	11.66	15.66	13.16	10.66	9.16	22.00	21.33	16.33	16.00
weight	+1.04	±1.04	±0.76	± 1.04	±1.25	±1.04	±0.76	±0.76	±0.50	±0.76	±0.57	±0.57	±2.0	±3.05	±3.51	±2.0
Shot dry weight	3.84	3.57	3.17	2.56	4.67	3.92	3.32	2.91	3.94	3.30	2.68	2.28	5.63	5.33	4.43	4.08
	±0.15	±0.010	±0.010	± 0.015	±0.01	±0.015	±0.01	±0.005	±0.02	±0.015	±0.01	±0.015	±1.52	±0.020	±0.32	±0.015
%MC root	66.41	66.81	66.79	66.85	66.98	67.71	67.20	67.41	66.94	66.53	66.98	66.87	66.90	66.86	66.66	68.62
%MC root*	196.77	201.36	201.16	201.74	202.90	203.96	204.91	206.74	202.54	198.85	202.87	201.87	202.13	201.78	200.00	218.77
%MC shoot	74.95	75.08	74.49	75.21	74.97	74.96	75.09	75.04	78.92	72.87	74.85	75.10	74.40	75.01	91.89	72.31
%MC shoot*	298.21	329.41	299.36	303.51	299.57	299.48	301.50	300.68	374.54	268.62	297.76	301.75	290.63	300.18	300.24	261.17

Values in mean, ± Standard deviation, length in cms and weight in gms.

Table 47.1(b) : Growth responses in terms of length, fresh and dry weight of root and shoot and per cent moisture content, both fresh and dry weight based in 80d old plants.

Attribute	*T. erecta*				*D. barbatus*				*D. caryophyllus*				*P. drummondi L.*			
	(pH of acid rain)															
	5.6	4.5	3.5	2.5	5.6	4.5	3.5	2.5	5.6	4.5	3.5	2.5	5.6	4.5	3.5	2.5
Root length	13.66	11.50	8.16	6.66	15.66	12.16	9.16	6.83	12.00	11.33	10.00	9.00	12.00	10.00	8.00	5.50
	±0.28	±0.86	±0.28	±0.28	±1.04	±0.57	±0.57	±0.76	±1.0	±0.58	±1.00	±1.00	±0.50	±0.50	±0.50	±1.00
Shoot length	21.83	18.33	16.50	15.50	21.50	18.50	15.83	12.00	16.50	14.83	10.00	9.00	16.16	13.33	11.50	8.66
	±0.96	±1.04	±0.50	±1.17	±0.50	±0_50	±0.76	±0.50	±0.50	±0.28	±1.00	±1.00	±1.52	±1.04	±0.50	±0.76
Root fresh weight	10.16	8.50	7.0	4.66	11.66	8.50	6.83	4.50	7.33	6.50	13.00	9.50	11.00	9.00	7.00	5.66
	±0.28	±0.50	±0.50	±0.76	±0.76	±1.00	±0.57	±1.00	±0.76	±0.50	±0.28	±0.50	±0.50	±0.50	±0.50	±0.76
Root dry weight	3.34	2.84	2.31	1.52	3.85	2.82	2.26	1.44	2.42	2.14	5.50	4.83	3.63	3.13	2.32	1.96
	±0.02	±0.02	±0.01	±0.02	±0.015	±0.015	±0.025	±0.01	±0.025	±0.02	±0.50	±0.76	±0.02	±0.02	±0.15	±0.015
Shoot fresh weight	21.50	18.66	18.16	15.83	13.16	11.50	10.00	8.00	19.66	16.66	13.66	12.00	14.50	11.66	8.33	5.16
	±0.50	±1.25	±2.02	±1.75	±0.28	±0.50	±0.50	±0.50	±1.52	±1.52	±1.52	±1.0	±0.50	±0.76	±1.25	±1.04
Shoot dry weight	539	4.16	4.53	3.95	3.29	2.85	2.53	2.10	4.93	4.17	13.66	12.00	3.63	2.93	2.06	1.29
	±0.01	±0.064	±0.01	±0.01	±0.20	±0.01	±0.01	±0.010	±0.20	±0.015	± 1.52	± 1.00	±0.011	±0.011	±0.020	±0.010
%MC root	67.12	66.58	67.00	67.38	66.98	66.82	66.91	68.00	67.38	67.07	67.09	67.04	67.00	65.22	66.85	67.13
%MC root*	204.19	199.29	203.03	206.57	202.85	201.41	202.21	212.51	205.78	203.73	203.86	203.82	203.03	187.53	201.72	204.30
%MC shoot	72.35	75.29	75.05	75.04	75.00	75.21	74.70	73.75	74.92	74.96	74.96	74.16	74.96	74.87	75.27	75.00
%MC shoot*	261.78	304.47	300.88	300.75	300.01	303.50	295.25	280.95	296.47	298.78	299.41	287.09	299.44	297.95	304.36	300.00

Values in mean, ± Standard deviation, length in cms and weight in gms.

Table 47.2(a) : Measurement of chlorophyll content, ascorbic acid and sulphur (mg/g fwt.) and leaf extract pH in s.a.r. treated leaves of different plants at the age of 80d.

Attribute	*C. roseus L.*						*C. officinalis L.*					
	pH of s.a.r.											
	5.6	4.5	3.5	2.5	CD 5%	CD 1%	5.6	4.5	3.5	2.5	CD 5%	CD 1%
Chlorophyll *a*	1.136	1.126*	1.102	1.078*	0.00	0.01	1.120	1.118*	1.112*	1.108*	0.01	0.02
Chlorophyll *b*	0.739	0.731*	0.708*	0.669**	0.01	0.02	0.802	0.790	0.762*	0.703*	0.01	0.02
Ascorbic acid	0.470	0.450	0.420*	0.410**	0.03	0.07	0.430	0.420*	0.410*	0.390*	0.02	0.04
Sulphur	3.980	4.370*	4.790*	5.070*	0.04	0.08	3.980	4.530*	4.890**	5.020*	0.03	0.07
Leaf extract pH	6.820	6.480	6.240*	6.210*	0.04	0.08	6.450	6.240*	6.130*	16.030**	0.04	0.08

Attribute	*C. indicum L.*						*D. pinnata 'Cav'*					
	pH of s.a.r.											
	5.6	4.5	3.5	2.5	CD 5%	CD 1%	5.6	4.5	3.5	2.5	CD 5%	CD 1%
Chlorophyll *a*	1.118	1.114*	1.109*	1.068*	0.00	0.01	1.119	1.112*	1.110*	1.090*	0.01	0.03
Chlorophyll *b*	0.776	0.772*	0.748*	0.723*	0.03	0.06	0.782	0.768*	0.748**	0.682**	0.01	0.02
Ascorbic acid	0.460	0.430	0.410*	0.410**	0.04	0.08	0.440	0.420*	0.410*	0.390*	0.02	0.04
Sulphur	3.390	3.540*	3.890**	4.210*	0.02	0.04	3.660	4.280*	4.820**	5.180**	0.03	0.07
Leaf extract pH	6.480	6.340	6.130	6.080*	0.03	0.06	6.620	6.480*	6.220**	16.020**	0.04	0.08

Values in mean, CD Critical difference, *Significant at 5% level, ** Significant at 1% level.

Table 47.2(b) : Measurement of chlorophyll content, ascorbic acid, sulphur (all in mg/g fwt.) and leaf extract pH in s.a.r. treated leaves of different plants at the age of 80d.

Attribute	*T. erecta L.*						*D. barbatus L.*					
	pH of s.a.r.											
	5.6	4.5	3.5	2.5	CD 5%	CD 1%	5.6	4.5	3.5	2.5	CD 5%	CD 1%
Chlorophyll *a*	0.852	0.839*	0.826*	0.808*	0.01	0.03	1.184	1.181*	1.176*	1.158**	0.01	0.02
Chlorophyll *b*	0.679	0.673*	0.659*	0.643	0.01	0.02	0.782	0.772*	0.752*	0.724*	0.01	0.02
Ascorbic acid	0.430	0.410	0.390*	0.380**	0.04	0.08	0.490	0.480*	0.460**	0.450*	0.04	0.08
Sulphur	4.120	4.310*	4.390**	4.610*	0.04	0.06	3.520	3.720*	3.890*	4.240*	0.05	0.10
Leaf extract pH	6.210	6.170*	6.130**	6.040**	0.02	0.08	6.540	6.340*	6.160*	6.080*	0.30	0.06

Attribute	*D. caryophyllus L.*						*P. drummondi L.*					
	(pH of s.a.r.)											
	5.6	4.5	3.5	2.5	CD 5%	CD 1%.	5.6	4.5	3.5	2.5	CD 5%	CD 1%
Chlorophyll *a*	1.134	1.129*	1.121	1.114*	0.01	0.02	1.259	1.254*	1.241*	1.237*	0.01	0.02
Chlorophyll *b*	0.748	0.738*	0.732*	0.719**	0.01	0.02	0.829	0.819*	0.812*	0.847*	0.00	0.01
Ascorbic acid	0.440	0.430*	0.420**	0.410*	0.02	0.04	0.460	0.440	0.410*	0.390**	0.02	0.04
Sulphur	3.120	3.140*	3.580*	3.620**	0.03	0.07	4.210	4.380*	4.720*	4.850*	0.04	0.08
Leaf extract pH	6.580	6.160*	6.080*	6.020*	0.04	0.08	6.630	6.340*	6.180**	6.070**	0.03	0.07

Values in mean, CD Critical difference, * Significant at 5% level, ** Significant at 1% level.

Table 47. 3 : Net primary productivity (NPP in g d^{-1}) in s.a.r. treated plants at 80d age.

Plant species	*NPP (gd^{-1}) pH of acid rain*			
	5.6	4.5	3.5	2.5
Catharanthus roseus L.	0.098	0.090	0.082	0.070
Calendula officinalis L.	0.092	0.080	0.069	0.056
Chrysanthemum indicum L.	0.088	0.074	0.059	0.048
Dahlia pinnata 'Cav'	0.117	0.115	0.101	0.090
Tagetes erecta L.	0.109	0.093	0.085	0.068
Dianthus barbatus L.	0.098	0.070	0.059	0.044
Dianthus caryophyllus L.	0.091	0.078	0.065	0.058
Phlox drummondi L.	0.090	0.075	0.054	0.039

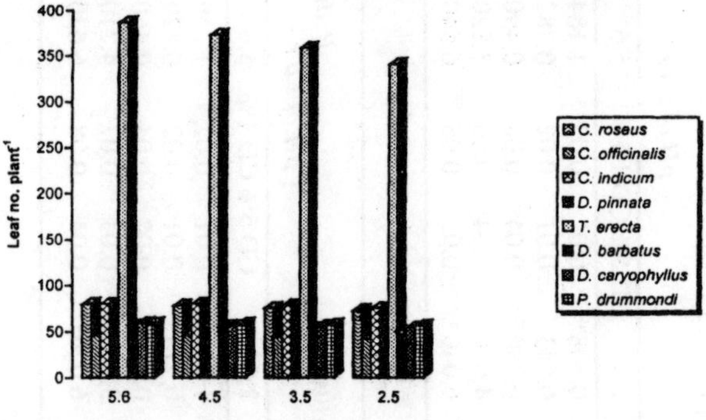

Fig. 47.2 : Leaf area $plant^{-1}$ in response to different pH of s.a.r. in different plant species.

the pollutant treated plants against the stress with reason that lesser leaf surface is likely to reduce the absorptive surface and thereby helping the plants to combat the deleterious effects of pollutant by lowering the gas flux into the leaves (Giridhar, 1981; Kumar, 1997).

A trend of reduction in leaf extract pH was also observed at all growth stages in all plants for all ranges of acidity. Low pH of acid rain proved more toxic to the green pigments in comparison to higher pH (Table 47.2 a &b). The pertubative effects were more pronounced on chlorophyll *a* and reduction of chlorophyll *a* was found to be more than chlorophyll *b* in exposed plants. Hence there was a decrease in chlorophyll *a* to *b* ratio which was recorded as 1.61, 1.58, 1.48, 1.59, 1.27, 1.59, 1.55 and 1.53 in *Catharanthus roseus, Calendula*

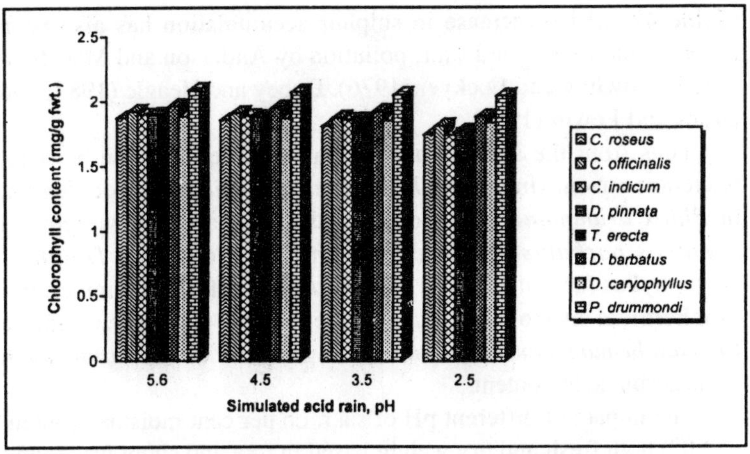

Fig. 47. 3 : Total chlorophyll content of leaves treated with different pH of s.a.r. in different plant species.

officinalis, Chrysanthemum indicum, Dahlia pinnata, Tagetes erecta, Dianthus barbatus, Dianthus caryophyllus and *Phlox drummondi* respectively at pH 2.5. The deleterious effect of s.a.r. on chlorophyll *a* and *b* resulted in decrease in total chlorophyll content (Fig. 47.3). Higher susceptibility of chlorophyll *a* to acid rain observed in the present study is in agreement with the reports of Khan and Khan (1994), Johnston *et al.* (1986) and Verma (1999).

Simulated acid rain treated plants, analyzed for ascorbic acid content, showed a loss in their content. The plants with lower ascorbic acid content showed higher sensitivity to s.a.r., i.e. the plants with higher concentrations of ascorbic acid were less susceptible to s.a.r. Thus there exists a correlation between the amount of ascorbic acid content and sensitivity to s.a.r. Reductions in ascorbic acid contents following SO_2 fumigation have also been noticed by Panda (1989), Shahare (1995) and Singh (2000). According to Chaudhary and Rao (1977) the tolerance to air pollution in plants is related with ascorbic acid levels. They concluded that higher the level of ascorbic acid, the greater was the tolerance (Table 47.2 a &b).

Simulated acid rain contrarily causes an increase in sulphur accumulation in plants studied with maximum effect in *Dianthus caryophyllus* and minimum in *Dahlia pinnata*. Plants which are more sensitive to acidic treatment, accumulate more sulphur

(Table 47.2a&b). Increase in sulphur accumulation has also been reported due to SO_2 and s.a.r. pollution by Anderson and Mansfield (1979), Cowling and Lockyer (1976), Dubey and Heagle (1987) and Evans and Lewin (1980).

Thus from the above study, it was observed that the per cent reduction was maximum in *Chrysanthemum indicum* for root length; in *Phlox drummondi* for shoot length, fresh and dry weight; in *Dianthus barbatus* for root fresh and dry weight; in *Dianthus* caryophyllus for leaf area and in *Dahlia pinnata* for leaf number. For biochemical attributes, per cent reduction was maximum in *Chrysanthemum indicum* for total chlorophyll and in *Phlox drummondi* for ascorbic acid content.

The impact of different pH of s.a.r. on per cent moisture content (%MC) both fresh and dry weight based in root and shoot was found quite different at different ages of different plants in different treatments. An increase in the dry weight based %MC in 80d old plant at pH 2.5 for root was observed in *Catharanthus roseus, Phlox drummondi, Calendula officinalis, Dahlia pinnata, Dianthus barbatus* and *Tagetes erecta;* for shoot in *Catharanthus roseus, Phlox drummondi, Calendula offcinalis* and *Tagetes erecta.* The decrease was noted for root in *Chrysanthemum indicum* and *Dianthus caryophyllus* and for shoot in *Chrysanthemum indicum, Dahlia pinnata, Dianthus caryophyllus* and in *Dianthus barbatus.*

Decrease in dry weight accumulation consequently reflects its impact on NPP (Table 47.3). These reductions were found to be increased in progressive manner with the increasing concentrations of acidity in all tested plants. Such types of observations have also been noticed by Ghouse and Khan (1984) and Narain and Singh (1984).

The increase in SRR and SWR and decrease in RWR indicate that the rate of translocation of photosynthate to the roots gets slowed down under the influence of air pollution which inhibit phloem loading system (Noyes, 1980 and Teh and Swanson, 1982). Such mode of phloem partitioning exists as these plants put greater priority to the growth of shoot than to roots under stress of pollution. Our results in this regard resemble to those of Mejstrick (1980), Shimizaki and Kiyashi (1980), Jensen (1981) and Gupta and Ghouse (1987).

The indication of these findings in terms of predicting the damages threatened by such pollutants on field grown crops is of much importance. Reductions in growth of all plants exposed to s.a.r. prepared them to reduce their potentiality for the yield.

REFERENCES

Adaros, G., Weigel, H.J. and Jager, H.J. (1988). Effects of sulphur dioxide and acid rain alone or in combination on growth and yield of broad bean plants. *New Phytol* 108: 67-74.

Anderson, L.S. and Mansfield, T.A. (1979). The effects of nitric oxide pollution on the growth of tomato. *Environ. Pollut. 20:* 113-21.

Amon, D.I. (1949). Copper enzyme in isolated chloroplasts, polyphenoloxidase in *Beta vulgaris. Plant Physiol.* 24: 1-15.

Chaudhary, C.S. and Rao, D.N. (1977). Study of some factors in plants controlling their susceptibility to SO_2 pollution. *Proc. Natl. Acad Sci. India. 46(B):* 236-41.

Cowling, D.W. and Lockyer, D.R. (1976). Growth of perennial rye grass exposed to low concentration of sulphur dioxide. *J. Exp. Bot.* 27: 411-14.

Dubey, D.T. and Heagle, A.S. (1987). The effects of simulated acid rain with and without ambient rain on the growth and yield of field grown soybeans. *Environ. Exp. Bot.* 27: 395-401.

Evans, L.H. and Lewin, K.F. (1980). Growth, development and yield response of pinto beans and soybeans to hydrogen ion concentrations of simulated acid rain. *Environ. Exp. Bot. 21:* 102-13.

Ghouse, A.K.M. and Khan, F.A. (1984). The effect of air pollution on the growth response of *Solanum nigrum* L. *Acta Bot Indica 12:* 93-97.

Giridhar, B.A. (1981). Relation between rates of absorptions of sulphur dioxide by plants and area of leaves. 68th *Indian Sci. Cong.* (abstract) pp. 158.

Gupta, M.C. and Ghouse, A.K.M. (1987). Effects of coal smoke pollutants from different sources on growth, chlorophyll contents, stem anatomy and cuticular traits of *Euphorbia hirta* L. *Geobios* 14: 221-22.

Gupta, V., Tomer, Y.S. and Prakash G (1993). Effects of SO_2 fumigation on *Lagenaria siceraria* L. *Acta Bot. Indiea 21:* 230-37.

Jacobson, J.S., Troiano, J.J., Hewller, L.I. and Osmetoski J. (1985). Influence of sulphate, nitrate, chlorate and chloride in simulated acidic rain on radish plants. *J. Environ. Qual 15:* 301-04.

Jensen, K.F. (1981). Ozone fumigation decreased the root carbohydrate content and dry weight of green ash seedlings. *Environ. Pollut. 26:* 147-52.

Johnston, J.W., Shriner, D.S. and Kinerley, C.K. (1986). The combined effect of simulated acid rain and ozone on injury chlorophyll and growth of radish. *Environ. Exp. Bot. 26:* 107-13.

Keelar, T. and Schwager, H. (1977). Air pollution and ascorbic acid. *Eur. Jour. Pathol.* 7: 338-50.

Khan, M.R. and Khan, M.W. (1991). Impact of air pollutants emanating from a thermal power plant on tomato. *J. Indian bot. Soc.* 70: 239-44.

Khan, M.R. and Khan, M.W. (1994). Effects of simulated acid rain and root-knot nematode on tomato. *Pl. Pathol. 43:* 41-50.

Kohut, R.J., Amuldson, R.G., Calavita L., Vam Luken, P.Z. and King, P. (1987). Effects of O_3 and SO_2 on yield of winter wheat. *Phytopathol.* 77: 71-74.

Kojiol, M.J. (1984). Interactions of gaseous with carbohydrate metabolism. In: *Gaseous Air Pollutants And Plant Metabolism* (eds.) M.J. Kojiol and F.R. Whatley. Butterworths London. pp. 251-73.

Kumar, V. (1997). Effects of simulated acid rain on *Zea mays* L. Ph.D. Thesis of C.C.S. Univ., Meerut.

Lee, J.J., Neely, G.E., Perrigan, S.C. and Grotheus, L.C. (1981). Effect of simulated acid rain on yield growth and foliar injury to several crops. *Environ. Exp. Bot. 21:* 171-85.

Linskens, H.F., Derks, F.H.M. and Schonenberg, Linders, G.F.J. (1988). The influence of artificial acid rain on the physiology and morphology of *Phaseolus vulgaris, Vicia faba* and *Spinacia oleracea* after separate spraying on shoots and soil. *Ange. Bot. 63:* 67-80.

Mejstrick, V. (1980). The influence of sulphur dioxide fumigation on the growth reduction of *N. tobaccum cv.* Samsum and C. *sativas cv.* Unicat. *Emviron. Pollut.* 21: 73-76.

Narain, S. and Singh, V. (1984). Effects of sulphur dioxide on *Triticale hexaploide. Nat. Acad Sci. Lett.* 7: 301-03.

Noyes, R.D. (1980). The comparative effects of sulphur dioxide on photosynthesis and translocation in bean. *Physiol. Pl. Pathol. 16:* 73-79.

Panda, S. (1989). Effects of environmental pollution on physiological aspects of plant. *Indian J. Appl, Biol. 4(1):* 55-58.

Patterson, G.D. (1958). Sulphur. In: *Calorimetric Determination of Nonmetals.* pp. 261-308. Interscience Publishers Inc., New York.

Shahare, C.B. (1995). Role of ascorbic acid as indicator of SO_2 pollution. *Geobios 22:* 34-38.

Shimazaki, K.I. and Kiyashi, S. (1980). Inhibition site of the electron transport system in lettuce chloroplasts by fumigation of leaves with sulphur dioxide. *Pl. Cell Physiol 21 :* 125-36.

Singh, N. and Rao, D.N. (1979). Studies on effects of sulphur dioxide on Alfa alfa plants especially under conditions of natural precipitations. *Ind J. Air Pollut. Con. 2:* 55-69.

Singh, N. (2000). Response of ornamental plants to air pollutant (sulphur dioxide). Ph.D. Thesis, C.C.S. Univ., Meerut.

Takemoto, B.K., Olszyk, D.M., Johnson, A.G. and Parada, C.R. (1988). Yield response of field grown crops to acidic fog and ambient ozone. *J. Environ. Qual. 17:* 192-97.

Teh, K.H. and Swanson, C.A. (1982). Sulphur dioxide inhibition of translocation in bean plants. *Plant Physiol. 69* : 88-92.

Verma, S.P. (1999). Simulated acid rain injury to germination and seedling growth of *Vigna unguiculata L.* Ph.D. Thesis, C.C.S. Univ., Meerut.

48

Study on Germination, Seedling Growth and Chlorophyll Content of Phaseolus Aureus CV. PS-16 under the Treatment of Paper Mill Effluent

S. Kumar, S. Lata and *R.M. Johri*

Introduction

Pollution has become alarming problem of threading the whole world. The problem of water pollution has been intensified due to increasing industrialization. Pollution due to industrial waste water is a serious problem and gaining a greater dimension day to day in India (Bhosale 1985). Usually the industrial waste water dumped off in to water bodies and make the water unfit for irrigation as well as drinking by increasing toxic elements. This creates an interest to explore the irrigation potential of the effluent of paper mill situated in Sahibabad industrial area of District Ghaziabad. The present studies were carried out under the effect of paper mill effluent on seed germination, seedling growth, and chlorophyll content of *Phaseolus aureus* CV. PS-16.

Materials and Methods

The effluent sample from paper mill situated in Sahibabad site IV industrial area of District Ghaziabad (UP) was collected from main discharging point of effluent and analysed for various heavy metals and a macronutrient. The effluent is yellowish brown in colour and disposed on land., hence it was decided to find the effect of the paper mill effluent on *Phaseolus auseus* CV. PS-16.

Seeds of proposed crop were procured from I.A.R.I. Pusa New Delhi to carry out the proposed study. For germination and seedling growth, experiments were conducted in petriplates, and for chlorophyll content in pot uniform and healthy seeds were selected was surface sterilized with 0.01 per cent mercuric chloride for two minutes and the seeds of *Phaseolus aureus* CV. PS-16 were kept for germination in sterilized petriplates with double layer of filter paper using various concentrations of effluent, namely 25 per cent 50 per cent, 75 per cent and 100 per cent. Distilled water was used as control in separate set. Experiment lasted for seven days. The emergence of radicle through seed was taken as the criterion for germination. For assessing the seedling third, fifth and seventh days old seedlings were selected and taken out from each petriplate. Radicle hypocotyl epicotyl length and germination percentage were determined. Fresh and dry weight of radicle, hypocotyl epicotyl and cotyledon were also recorded.

Chlorophyll estimation studies were carried out on *Phaseolus aureus* Roxb. CV.PS-16 plants treated with effluent. The leaves of plants were sampled at 15th, 30th and 45th day from radicle emergence. 600mg fresh leaves were taken from every sample and homogenized with 80 per cent acetone. The oplical density of the extract was at 645 nm and 665 nm on Data Processor Hitachi 150-20 spectrophotometer. The amount of cholorophyll a, chlorophyll b and chlorophyll a+b were calculated, according to the formula of Arnon (1949).

Result and Discussion

The different parameters like germination percentage, radicle length, hypcotyl length, eplcotyl length, fresh and dry weight of seedling of *Phaseolus aureus* were selected to investigate the effect of the paper mill effluent. Fresh and dry weights of the cotyledons were also considered. It was observed that germination of seeds was maximum in diluted effluent (up to 25 per cent) over pure water (control) and decreased with increasing concentration of effluent (Table 48.2) in pure effluent (100 per cent). The germination percentage decreased tremendously. This is in agreement with the result of Hari Om *et al.* (1994) in the case of okra.

Table 48.1 : Metals detected in paper mill effluent

Metals	*Concentration*
Calcium	23.3 ppm
Iron	1.01 ppm
Zinc	0.19 ppm
Chromium	Nil
Copper	0.15 ppm
Cadmium	0.01 ppm
Lead	0.03 ppm
Mercury	0.65 ppm
Nickel	2.4 ppm

Table 48.2 : Effect of paper factory effluent on the germination percentage of *Phaselous aureus CV. PS*-16.

Concentration of effluent	*Germination percentage*
Control	90
25%	97
50%	89
75%	74
100%	70

The data on effect of seedlings growth of 3rd, 5th and 7th days old seedling revealed the variation with respect to different concentration of the effluent. There was a gradual increase in the radicle, hypocolyl and epicotyl lengths of *phaseolus aureus* CV.PS-16 seedlings with the increase of dilution of the effluent up to 25 per cent concentrations.

Fresh and dry weight of the radicle, hypocolyl and epicotly have been reported in (Table 48.3) which indicates that the growth of the plant, i.e. radicle, hypocotryl and epicotyl increased from control to the 25 per cent effluent. There findings are in also coordination with our previous observations in this variety of *Phaseolus* under the impact of steel factory waste water.

The chemical analysis of paper mill effluent revealed the presence of some heavy metals like Zn, Fe, Cr, Cd, Pb, Cu, Ni, Hg, etc. and a plant nutrient Ca (Table 48.1), Arora *et al.* (1985) have also reported that waste water contain significant amounts of the plant nutrients, potentially hazardous elements. The results of the present study

Table 48.3 : Effect of paper mill effluent on the fresh and dry weight of seedling and Cotylodon of *Phareolus aureus* CV. PS-16

Concen-traion of effluent	*Days of emergence*								
	Fresh weight (Mg, ±SD)			*Dry weight (mg, ±SD)*			*Length (cm., ± SD)*		
	3	*5*	*7*	*3*	*5*	*7*	*3*	*5*	*7*
					Radicle				
Control	0.75 ±0.69**	11.11 ±0.63**	11.80 ±0.70**	1.10 ±0.09*	1.85 ±0.53*	1.80 ±0.56**	2.04 ±0.70*	3.65 ±0.39**	4.16 ±0.31*
25%	17.50 ±0.66**	11.63 ±0.74*	15.80 ±0.83**	1.68 ±0.43**	2.04 ±0.31*	1.89 ±0.56*	3.57 ±0.08**	3.28 ±0.51**	5.30 ±0.97**
50%	11.25 ±0.50*	9.90 ±0.85**	11.20 ±0.64*	1.20 ±0.13**	1.53* ±0.37**	1.59 ±0.37**	2.93 ±0.87**	3.03 ±0.59*	3.25 ±0.63**
75%	9.20 ±0.47**	9.33 ±0.66**	10.20 ±0.63**	0.98 ±0.05**	1.4 ±0.31**	1.40 ±0.27**	1.76 ±0.24**	2.03 ±0.37**	3.11 ±0.40**
100%	6.25 ±0.29*	6.42 ±0.68*	9.80 ±0.58**	0.88 ±0.98**	1.47 ±0.30*	0.90 ±0.03**	1.67 ±0.63**	1.90 ±0.54*	2.32 ±0.40*
					Hypocotyl				
Control	83.00 ±1.92*	122.40 ±4.60*	124.00 ±4.30*	6.43 ±0.68**	8.65 ±0.51*	7.48 ±0.39**	3.83 ±0.89*	7.81 ±0.57*	9.14 ±0.66*
25%	99.75 ±5.74**	124.44 ±4.79**	135.00 ±r.70**	8.02 ±0.63**	10.03 ±0.70*	8.21 ±0.46*	4.15 ±0.35**	9.15 ±0.41*	9.50 ±0.53**
50%	60.00 ±3.22**	104.00 ±2.70*	117.50 ±4.70**	5.91 ±0.60**	8.27 ±0.65*	6.46 ±0.38**	2.81 ±0.60*	6.81 ±0.58**	9.38 ±0.59**
75%	55.62 ±3.18**	91.42 ±2.60**	111.00 ±3.22**	5.46 ±0.38**	8.20 ±0.64**	6.43 ±0.37*	2.73 ±0.61**	6.28 ±0.97**	9.28 ±0.85**
100%	43.75 ±2.10**	80.20 ±3.21**	107.00 ±4.42**	4.61 ±0.38**	6.92 ±0.58*	5.16 ±0.33**	2.22 ±0.26**	4.73 ±0.55**	9.17 ±0.96*
					Epicotyl				
Control	7.75 ±0.47**	9.40 ±0.68*	47.11 ±1.26*	1.21 ±0.13*	1.98 ±0.25*	4.12 ±0.26*	0.81 ±0.27*	1.80 ±0.33*	6.62 ±0.52*
25%	9.37 ±0.67*	10.00 ±0.57**	59.20 2.53**	1.50 ±0.16**	2.15 ±0.21*	4.56 ±0.47**	1.11 ±0.16**	1.82 ±0.18**	8.15 ±0.72*
50%	4.00 ±0.07**	7.80 ±0.53**	41.42 ±1.09**	1.02 ±0.01*	1.50 ±0.31*	4.04 ±0.34**	0.75 ±0.17*	1.37 ±0.21**	6.12 ±0.35**
75%	3.62 ±0.50**	7.40 ±0.25*	38.00 ±1.73**	0.77 ±0.06**	1.48 ±0.30*	3.58 ±0.37**	0.68 ±0.16**	1.35 ±0.20**	5.31 ±0.46**
100%	2.75 ±0.47**	6.20 ±0.52**	29.77 ±0.79**	0.66 ±0.06**	1.32 ±0.23*	3.18 ±0.22**	0.38 ±0.07*	1.12 ±0.33**	5.25 ±0.36**
					Cotyledon				
Control	26.00 ±2.03*	14.00 ±0.67*	9.80 ±0.64**	9.61 ±0.49**	2.59 ±0.06*	1.33 ±0.20*			
25%	21.25 ±0.70**	12.5 ±0.65**	9.00 ±0.42**	9.16 ±0.22**	2.25 ±0.16*	1.05 ±0.19**			
50%	27.75 ±0.79**	18.33 ±0.66**	12.50 ±0.63**	12.25 ±0.65**	3.31 ±0.27*	1.43 ±0.31*			
75%	3.75 ±1.95*	20.66 ±0.75*	14.00 ±0.54**	12.26 ±0.65*	4.01 ±0.50*	1.66 ±0.10*			
100%	34.00 ±1.08**	21.66 ±0.73**	16.10 ±0.63**	17.63 ±0.74**	5.11 ±0.94*	20.3 ±0.10*			

revealed that the highest concentration of the effluent inhibited both seed germination and seedling growth. It may be due to the presence of nutrient (Ca) in the effluent as observed by Arora *et al.* (1985). Present observation of increase and decrease in yield parameters also matches with the similar findings of Shinde et al. (1988) and Srivastava *et al.* (1988). The samples from the different paper factories showed considerable amount of nitrate, nitrogen, total phenols, carbonates, chlorides, suspended and dissolved solids (Rajannan and Oblisami, 1979, Billings de Hass, 1971, and Bowonder and Chandamanas 1973). The high amount of total suspended and dissolved solids and sulphates might impart toxicity to the effluent (Swaminathan and Vaidhessvaran, 1991). The reduction in growth of seedling (Table 48.3) might be due to presence of excess amount of soluble salts like calcium and magnesium and solid materials effluents as they cause injuries to plant system (Israelsean and Hansel, 1962). Thukral and Kaur (1987) have already found the adverse effect of elements such as copper and zinc on seedling growth. As these are widely known as constituents of many metallo-enzymes and played a major role in disclosing the physiological significances (Nagaraja, 1988).

The fresh and dry weight of cotyledon at 25 per cent effluent concentration decreased considerably however it decreased with increase in the effluent concentration from 25 per cent onward.

The growth reduction in the plant system is undoubtedly complex including toxic effects from excess quantities of micronutrients, heavy metals and salt toxicity (Dolar *et al.*, 1972). However diluted effluent (25 per cent) enhanced the seed germination as well as seedling growth this might be due to the decrease in the concentration of various chemicals in the effluents. These observations are in conformity with those of Sahai *et al.* (1983), who observed better seed germination and seedling growth in lower concentrations of the effluent.

There is also increase in content of chlorophyll 'a', chlorophyll 'b' and chlorophyll a+b on 15th, 30th and 45th days of radicle emergence at 25 per cent effluent concentration and further chlorophyll content decreased with increasing the effluent concentration. The maximum inhibition was found at 100 per cent effluent concentration.

From this study it can be concluded that the paper mill effluent after proper dilution (up to 25%) may safely be used for irrigation purpose as a liquid fertilizer but long term effects of this effluent may cause adverse effects because of the bio-magnification of metals.

REFERENCES

Arnon, D.I. (1949) Copper enzymes in isolated chlorophlasts polyphenyloxidase in *Beta vulgaris* Plant physical. 24, 1-15.

Arora, B.R., A.S. Azad, Bijay Singh and, G.S. Sekhon, (1985) Pollution potential of municipal waste water of Ludhiana, Punjab, India, *J. ecol.* 12(1) : 1-7.

Bhosale, C.J. (1985) *Effect of water pollution on plants. In Current Pollution Researches in India* (eds. R.K. Trivedy and P.K. Goel) Environmental publication Karad, India.

Billings, R.M. and, G.G. de Hass, (1971) Pollution control in the pulp and paper industry. In *industrial pollution control Hand-book* (ed.) H.F. Lund, McGraw Hill Book Co., New York, pp. 18, 1-18, 28.

Bowonder, B. and M. Chandrmanas, (1973) Water pollution and environment, *India Chem. Manufact.* 10, 9-19.

Dolar, S.G., J.R. Boyle, and A.D. Keenay (1972) Paper mill sludge disposal on soil. Effects on the yield and mineral nutrition of Oats (Avena sativa L.) J. *Environ. Qual.* 1, 405-09.

Hariom, Nepal Singh and S. Arya Manjank (1994) Combined effect of wastes of distillery and sugar mill on seed germination seedling growth and biomass of Okra. *J. Environ. Biol*, 15(3), 171-75.

Israelsen, O.W. and Hansen, V.E. (1962) *Irrigation : Principle and practices,* John Wiley and Cons, Inc. New York.

Nagaraja, T.G. (1998) Mineral composition of Cassia fistual L. leaves infected with saracinella cassiae *Ad. Plant Sci.* 11(1), 291-93.

Nandita Shukla and J.K. Moitra (1995) Effect of integrated steel plant effluent on growth parameters of selected pulses and cereals. *J. Environ. Biol.* 16(1), 71-73.

Rajannan, G. and G. Oblisami (1979) Effect of paper factory effluent on soil and crop plants, *Indian J. Environ. Hlth.* 21 (2), 120-30.

Sahai, R.S., Jabeen, and P.K. Saxena 1983 Effect of distillery effluent on seed germination, seedling growth and pigment content of rice, *Indian J. Ecol.*, 10(1), 7-10.

Shinde, D.S., R.K. Trivedy and S.D. Khetavkar (1988) Effect of pulp, paper and sugar factory waste on germination of gram. *Pollu. Res* (3-4) % 117-122.

Srivastava, N. Neelam and R. Sahai (1988) Ecophysiological response of certain crop plant of eastern Uttar Pradesh in industrial effluent. *Ind. J. Environ. Sci.* 14 : 229-43.

Swaminathan, K. and P. Vaidheeswaram (1991) Effect of dyeing factory effluents on seed germination and seedling development of ground nut (Arachis hypogea). *J. Environ. Biol.* 12(4), 353-58.

Thukral, A.K. and Parminder Kaur (1987) Effect of some trace elements of polluted water on the germination of Cyamposis tetragonoloba Taub. *Indian J. Ecol.*, 14(2), 185-88.

49

Implications of Commercial Use of Biotechnology in Vegetable Crops

Suruchi Tyagi

Biotechnology are contributing increasingly to the improvement of agricultural crops, especially with the introduction of new desirable traits targetted to meet the consumer requirement. Consumer trends point towards increasing demand for quality, novelty, consistency and convenience among horticultural commodities. In agricultural, targetted traits include aroma, flavours, texture, pigments, resistance to disease and insect resistance, nutritional quality and shelf life. Plant breeding had been traditionally aimed at enhancing yield and not consumer oriented crop traits. Rapid advances in the new technologies especially molecular and cellular biology can make it possible to bring about profound genetic changes in crop varieties.

Present Status of Commercialisation of Agricultural Crops

Today plant tissue culture technology has been globally commercialized. Many companies in India and abroad are now producing large quantity of planting materials especially of fruits, vegetables and ornamental crops. Over 800 commercial companies are engaged in such activities all over the world. Their number is increasing day-by-day. Annual plantlet production is approximately 513 million plantlets. In India, the commercialization of tissue culture began in 1987, when a number of private sector organizations looked at this technology as a means of increasing their productivity.

The first commercial tissue culture unit was set up by M/s A .V. Thomas & Co. (AVT) at Chohin, which was soon followed by Indo American Hybrid Seeds at Bangalore. Subsequently private companies like the Hindustan Lever, Tata Tea, Unicon Biotech, Nath Seeds, RPG Enterprises, Indian Tobacco Company, Maharashtra Hybrid Seed, Harrison Malayalam and Southern Petrochemical Industries Corporation (SPIC), took up a large scale production of plants through tissue culture. Public funded organization like—National Chemical Laboratory (NCL), Pune, and another non-profiting organization, Tata Energy Research Institute (TERI), New Delhi, with pilot scale facility are also engaged in a large scale micropropagation of elite forest tree species with a production potential of million of plantlets year. The Department of Biotechnology of the Ministry of Science and Technology, the nodal department in the government, is promoting micropropagation of a good number of commercial crops in the country.

There were only 5 field trials of transgenic crops in 1986. During the last 10 years, 3647 field trials were conducted in 8 crops (corn, canola, potato, tomato, soybean, cotton, tobacco and squash/melon), including 3320 in industrialized countries, mostly in USA and Canada, 291 in developing countries, mainly in Latin America and 36 in Eastern Europe/Russia.

Besides, several field trials of GMOs are being conducted for breeding programmes in these countries. The most frequent categories of field trials of transgenic crops during 1986-95 in the world, USA and developing countries include herbicide tolerance, insect reistance, disease resistance (viral and fungal) and product quality. Global acreage under transgenic crops was about 30 million acres in 1997 (Vasil, 1998), the acreage being maximum in USA, 20.1 million acres (Medley, 1998). According to Mr. Graham Blight, President of the International Federation of Agricultural Producers, about US$ 18 billion worth of crops are grown now in the world from genetically engineered products.

Among vegetable crops tomato is the most vulnerable crop for biotechnology. The main emphasis in tomato was to develop transgenics with delayed fruit ripening and virus resistance. During 1986-95, 353 field trials of transgenic tomato were conducted worldwide, including USA (268), developing countries (33) and other countries (52), including two trials in Australia. Among the developing countries China had 9 trials and Thailand 2 trials and a few trials in

Mexico (James, 1996). Till February 1998, in USA alone there were 377 field testings of tomato transgenics (Medley, 1998). A few field trials of tomato transgenics have also been conducted in Europe (James, 1996).

Field trials were fewer (92) in melon/squash during the 10-year period (1986-1995) worldwide. Of which 82 were in USA and 10 in other countries. The USA had 16 field trials in cucumber and 8 in lettuce (Table 49.1). The data of USA compiled till February 1998 showed 126 field trials in melon/squash (Medley, 1998). There were 4 field trials of transgenic pepper in China, and one trial each in asparagus and broccoli in New Zealand. Transgenic eggplant field trials have been conducted in Eastern Europe/Russia (James, 1996).

Table 49.1 : Number of Transgenic vegetable crop field trials worldwide (1986-1995)

Country	*Tomato*	*Squash/ Melon*	*Cucumber*	*Lettuce*	*Asparagas*	*Broccoli*	*Pepper*
Worldwide	353	92	16	8	1	1	4
USA	268 (377)*	82 (126)*	16	8	–	–	–
Australia	2	2	–	–	–	–	–
Japan	6	1	–	–	–	–	–
China	9	–	–	–	–	–	–
Thailand	2	–	–	–	–	–	4
New Zealand	–	–	–	–	1	1	–

Source : James (1996), *Medley (1998).

Five or less number of field trials were conducted in USA during 1987-1997 in eggplant, onion, pepper and watermelon (Medley, 1998).

Commercial Releases of Transgenic Vegetable Crops

China was the first country to introduce the viral resistant tobacco and tomato for commercial use in the early 1990s. Subsequently in 1994, Calgene Inc., USA, introduced its Flaur Savr^{TM} delayed ripening tomato for commercial use.

By May 1997, 8 transgenic crops involving 10 traits and 21 processes were approved for commercial use in USA (Ghosh, 1997).

Of these, 4 crops involving 8 processes belonged to horiticulture (Table 49.2). It is expected that several approvals might have been accorded in China, Japan, Australia, Argentina, Mexico and certain European countries. Consumer response to transgenic plant products has now been tested with the commercial release of improved varieties of many crops. These releases were accompanied with dissemination of information on transgenic plants. The consumers have responded positively to quality or price advantage. The demand has outstripped supply in the first season in many cases. The area under transgenic crops has been increasing steadily during last 3-4 years. The area under transgenic crops was about 1.6 million hectares in 1996 (Ghosh, 1997), 75 per cent of this land being in USA followed by Canada (20 per cent) and other countries (5 per cent).

Table 49.2 : Transgenic vegetable crops released for commercialization

Crop	*Altered trait and gene used*	*Company*	*Year and Product*
Tomato	Delayed ripening (polygalac-turonase gene Sequence reversed to minimize its expression By antisense technology	Calgene Inc.	1994, Flavr Savr ™
	Delayed ripening (aminocyclopropane carboxylic and acid synthase gene from *Pseudomonas chlorphis 6G5 strain)*	DNA Plant Tech.	1995, Endless Summer™
	Delayed ripening (aminocyclopropane Carboxylic and deaminase gence from *Pseudomonas chlorphis 6G5 strain)*	Monsanto Co.	1995
	Thicker skin and altered pectin content For paste consistency (fragment of Polygalacturonase gene from tomato)	Zeneca/ Petoseed	1995
Potato	Resistance to Colorado potato beetle (CryIII A gene from Bacillus) Thuringiensis tenebrionis	Monsanto Co.	1995, New Leaf™
	Insect resistance [Modified CryIII (A) gene]	Monsanto Co.	1996
Squash	Virus resistance (CP genes of watermelon mosaic virus 2 and Zucchini yellow mosaic virus)	Asgrow Seed Co.	1995, Freedom II™
Papaya	Virus resistance (CP genes of PRSVHA 5-1 from Hawaii)	Cornell Univ., USA	1997

Source : James (1996), Medley (1998), Millis (1998), and Cheng (1998).

Besides tomato, potato, papaya and squash, there is no other vegetable crop in which transgenics have been released for commercial use, though several transgenics in cucumber, melon, lettuce, asparagus, broccoli, eggplant, onion and pepper are still being tested in field trials.

Current Biotechnology Research in Vegetable Crops

Many interesting and useful researches are being carried out in laboratories abroad which have great potential for commercial application. A few such examples are mentioned here. Production haploids by unpollinated ovules in onion (Muren, 1989) and by anther culture in broccoli, cabbage and other Brassica vegetables (Keller and Armstrong, 1983; Orton and Browers, 1985) are being used for hybrid seed production in these crops. In cabbage and cauliflower, self-incompatible lines are maintained by dihaploid technique and utilized for hybrid seed production by many seed companies. Crisp and Tapsell (1993) reported that in the United Kingdom attempt was made to produce F_1 hybrid varieties by anther culture which perhaps did not have commercial viability. Somatic hybridization between radish and cabbage using radish cytoplasm in protoplast fusion had shown much promise in development of cytoplasm-gene type of male sterility for production of hybrid seeds.

At present efforts are being made to overcome the problem of chlorosis in leaves of young seedlings (Bannerot *et al.*, 1974; Robertson *et al.*, 1987; Kagami *et al.*, 1990). Multiplication of seedless watermelon (triploid) by *in vitro* spropagation (Song *et al.*, 1988) may have commercial use. Triploid seedlings of seedless watermelon having less vigour do not successfully establish on transplanting in field. The production of 'synthetic seeds' in carrot will be useful in breeding programmes (Kamada *et al.*, 1989).

Greater emphasis may be on application of biotechnology in vegetable breeding, genetic engineering, transgenics, gene coding, selectable DNA markers, gene mapping and linkages with Qts, RFLP, AFLP, RPAD, STS etc., for finger printing of genotypes, varietal identification, phytogenetic relationship, analysis of genome organization, germplasm introgression, map based cloning and plant pathogen interactions. The time and cost of RFLP assays is a limiting factor in its commercial application. The development of breeder friendly' diagnostic kits for RFLP analysis seems to have commercial feasibility (De verna and Alpert, 1990). Mapping QTL by RFLP analysis in tomato has been undertaken by Jones and St. Clair (1997) for resistance to late blight. Resistance to yellow leaf curl virus in tomato has received attention of biotechnologists in recent years in USA (Lindhout, 1997). Similarly, in tomato work is in progress on spotted wilt virus (Van Staden, 1997) and tomato mottle virus (Stevens, 1997).

Future Prospects

Mr. Graham Blight, President of the International Federation of Agricultural Producers at the World Farmers Congress, held during 1998 in Manila (The Philippines) said that it was possible to develop every single Vegetable variety through genetic engineering in the next 5-10 years. "All variety of vegetables can be genetically engineered to produce better quality and bigger crops, bringing hope to a growing population". The fast developing biotechnology can bring gene revolution in many crops including vegetables, bringing improvement in productivity, product quality, resistance to biotic and abiotic stresses, better keeping quality and reducing post-harvest losses. It will be beneficial in ensuring nutritional security in developing countries.

REFERENCES

Barmerot, H.O., Bouliadard, L., Cauderon, Y. and Tempe, T., (1974) Cytoplasmic male sterility transfer from Raphanus to Brassica. Eucarpia Cruciferae Conference, Scottish Horiticultural Research Institute, Mylnefield, Dundee, Scotland, 52,1974.

Blight, Graham, (1998) *World Farmers Congress. Genes : Dicey, Manila*, Philippines. (Reuters news).

Cheng, Jingen, (1998) Regulations and administration for genetically manipulated agrobio products in China. Asean Workshop on Regulations for Agricultural Products derived from Biotechnology, held during 12 April, 1998, at Singapore.

Clark, D.L., Stevens, M.R. and Scott, J.W., (1997) RAPD polymorphisms associated with ToMoV resistance in tomato derived from L. chilense Durn, Eucarpia tomato 97, XIII Meeting of the Eucarpia Tomato Working Group, 19-23 January, 1997, at Jerusalem, Israel.

Crisp, P. and Tapsell, C.R., (1993) Cauliflower Brassica oleracea L., In : *Genetic*

Improvement of Vegetable Crops, pp.157-78. Kalloo, G. and Bergh, B.O. (Eds.), Pergamon Press.

De Verma, J.W. and Alpert, K.B., (1990) RFLP Technology, Horticultural Biotechnology pp.247-59 Bennett, A.B. and O'Neill, S.D. (Eds.), Wiley-Liss,Inc.

Fraley, R.T., (1994) The contributions of plant biotechnology to agriculture in the coming decades, In"Biosafety for sustainable agriculture: sharing biotechnology regulatory experiences of the Western Hemisphere" pp.3-28, Krattiger, A.F. and Rosemarin, A. (Eds.). ISAA : Ithaca & S.E.I.: Stockholm.

Ghosh, P.K. (1997) Biotechnolgy and Biosafety Issues in Agriculture. (in) *National Conference on Transgenics*, Tissue Culture, Floriculture, 20 October, 1997. All India Biotech Association, New Delhi. pp.2-13.

James, Clive, Teng. P.S. and Nutter, F. W., (1991) Estimated losses of crops from plant pathogens, In: *CRC Handbook of Pest Management*, VoI.I., pp. 15-50, Pimentel, D. (Ed.). CRC Press, Boca Raton, USA.

James, Clive, (1996) Progress of genetic engineering in field and horticultural crops. Technical Report No. 12, pp 23, Asia and Pacific Seed Association.

Jones Elizebeth and St. Clair, D. (1997) Progress towards QTL mapping of late blight resistance in a wild accession of Tomato Eucarpia Tomato 97., XIII Meeting of the Eucarpia Tomato Working Group held during 19-23 January, 1997 at Jerusalem, Israel.

Kagami, T., Akamatsu, T. and Shiga, T., (1990) Effective transfer of Ogura-type cytoplasmic male sterile traits from B. campestris or R. sativus to B. Oleracea by protoplast fusion, Proceedings of the 6th Crucifer Genetics Workshop, McGerson, J.R., Kresovich, S., and Dwyer, S.G.(Eds.), Cornell University, N.Y.

Kamada, H., Kobayashi, K., Kiyosure, T. and Harada, H, 1989. Stress induced somatic embryogenesis in carrot and its application to synthetic seed production. *Cell Div. Biol.* 25 : 1163.

Keller, W.A. and Armstrong, K.C. (1983) Production of haploids via anther culture in Brassica oleracea vac. italics cultivar Green Mountain, Euphytica 32: 1510.

Lindhout, P., (1997) Digging resistance genes in the Lycopersicon genus. Eucarpia *Tomato* 97, XII. Meeting of the Eucarpia Tomato Working Group held during 19-23 January, 1997 at Jerusalem, Israel.

Medley, T.L., (1998) U.S. regulatory oversight for the safe development and commercialization of plant biotechnology. Asean Workshop on Regulations for Agricultural Products derived form Biotechnology, held during 1-2 April, 1998 at Singapore.

Millis, Nancy, F., (1998) Surveillance system for gene technology in Australia. Asean Workshop on Regulations for Agricultural Products derived from Biotechnology, Held during 1-2, April, 1998 at Singapore.

Muren, E.C., (1989) Haploid plant induction from unpollinated ovaries in onion. *Hort. Science* 24 : 833.

Orton, T.J. and Browers, M.A., (1985) Segregation of genetic markers among plants regenerated from cultured anthers of broccoli (Brassica oleracea var. italica), *Theor. Appl. Genet.* 69 : 637.

Robertson, D., Palmer, J.D., Earle, E.D. and Mutschler. M.A., 1987. Analysis of organelle genomes in somatic hybrid derived from cytoplasmic male sterile Brassica oleracea and attrazine-resistant B. campestris. *Theor. Appl. Genet.* 74 : 303

Song, P.l., Pen, W.Z. and Yang, Y.H., 1988. In vitro propagation of seedless watermelon. Acta Scientaiarum naturalium Universities Narmalis Hunanesis 11 : 340.

Stevens, M. Allen. (1986) The future of the field crop. (in) the tomato Crop—A Scientific Basis for Improvement pp. 562-79. Atherton, J.G. and Rudich, J. (Eds.), Chapman and Hall Ltd., London, New York.

Van Station, D. and Relief, A.E., (1997) DNA markers for the tomato spotted wilt virus resistance gene (Sw-5). Eucarpia Tomato 97, MR Meeting of the Eucarpia Tomato Working Group, held during 19-23 January, 1997 at Jerusalem. Israel.

Vasil, I.K. (1998) Biotechnology and food security for the 21st century. A real world perspective. *Nature Biotechnology* 16, May 1998: pp.399-400.

SECTION-VI

SOCIO-CULTURAL ASPECTS

50

Population Development and Food Security

Aslam Mehmood

Introduction

Recent pace of population growth is unprecedented in the whole history of mankind. Till the beginning of 1880s the number of human population could reach only one billion (1000 million) and until 1960 it could reach 3 billion mark only. It has now reached 6 billion mark. Today, the world gains 1 billion people in every 12 years or around 80 million people each year, equivalent to 220,000 people a day (Engelman, 1997, p.11).

More than 90 per cent growth in the world population takes place in developing countries of Asia, Africa and Latin America. Although natural increase rate is smaller in Asia compared to Africa and Latin America, the vast continent has 3/4th of the world population and thus adds more people to the population of the world than any other continent. Population density is also found to be highest in Asia. It is more than 108 persons per square kilometre on the average. Its average value is 23 persons per square kilometre in Latin America, 24 persons per square kilometre in Africa and 14 persons per square kilometre in North America (UN, 1996).

The population pressure is likely to be the main concern of the human societies as even if present population could live safely, with the existing the Earth. As due to past experience of fertility an unusually higher proportion of population consists of young people in the child bearing age and these people contribute substantial population momentum of growth which is a critical demographic factor today and will continue for sometime to come also.

Population and Resources

In the context of economic growth of western Europe and North America, Kuznet (1966) defined modern economic growth as a sustained increase in population. Population growth of industrialized Europe or Western Europe at that time was however slow. It seldom exceeded 1.0 per cent per annual as compared to the rapid population growth of developing countries during the post-Second World War. Towards the beginning of the second half of the twentieth century mortality in most of the developing countries started falling down dramatically without any accompanying fall in fertility. As a result population growth rates in most of the developing countries registered unprecedented increase ranging from 2.0 to 4.0 per cent per annum. The consequences of population growing faster than the means of subsistence were initially outlined by Malthus (1959) in his famous essay on population. The means of subsistence, Malthus claimed, grew only at arithmetic rate, whereas populations tended to grow at geometric faster than the arithmetic rate. When the imbalance between growth in the means of subsistence and in population becomes too great, factors such as hunger, epidemic diseases and war—which Malthus called "positive checks"— would operate to increase the deaths and reduce the population to a level compatible with the means of subsistence.

Later he added "preventive checks" also related to births such as late marriage, and other means of family planning.

Another important theory explaining rapid population growth in developing countries is demographic transition theory (Notestein, 1945). As per demographic transition every country passes through three stages of demographic transition. In the first stage both birth rates and death rates are high and growth rate of population remain low. As economy progresses, education spreads, standards of living improve and medical and health facilities improve the death rates go down. At this second stage of the demographic transition the difference between birth rate and death rate is high and population grows with a very high rate. However, in developed countries this stage does not remain for long and with a lag of few years fertility also starts declining and the third stage of demographic transition takes place. At this stage both fertility and mortality decline and population growth stabilizes to a low level. In agrarian economies of the third world, the factors responsible for decline in fertility remain absent and it does not decline and populations remain at second stage of demographic transition of high population growth rate. The theory has several drawbacks also (Kirk, 1996). Although there is no unanimity among the scientists over the precise impact of population growth on development as tracing the cause and effect in this case is not direct. The cumulative evidence, however, is strong enough to show that current rates of population growth pose significant risk to human well being and are legitimate concern for all nations. In a short span of time we have noticed that human population has become a force on earth that rivals nature. The reasons for this are complex and are linked to changes of not only in human population but also in technology, consumption patterns, unequal distribution of wealth etc.

Population and Economic Development

Economic development according to the theory of Demographic Transition has the effect of bringing about a reduction in the death rates. Such reduction in death rate can take place either due to greater regularity in food supply (Coale and Hoover, 1958) and improvement in law and order etc. or due to advancement in medical technology or due to both (UN, 1953). Both of these factors of death rate reduction are rooted in economic development of the society. In developing countries, however, the rapid growth of population may make it difficult to accomplish these goals of economic and social changes, which are necessary for fertility reduction (Coale and Hoover, 1953, p.17). There appears to be a vicious circle between population change and development. In developing countries higher population growth will slow down social and economic development and as a consequence these countries are unable to reduce their population growth rate. The process continues even today and will be highlighted if we compare the population growth and socio-economic development in developed and developing countries of today.

To start with the growth rate of population in developed and developing countries of the world is given in Table 50.1.

Table 50.1 : Population Growth Rate

Region	*Period*			
	1950-55	1990-95	1995-2000	2045-50
More Developed	1.21	0.40	0.26*	-0.23*
Less Developed	2.05	1.77	1.65*	0.55*

* Projected
Source: World Population Prospects the 1996 version, UN, 1998, p.10.

From Table 50.1 it is quite clear that the population/growth rate of more developed regions of world is consistently lower than the less developed region and is likely to show higher reduction in latter part of this millennium and is likely to show higher reduction. The result of such a lower population growth rate in developed countries in comparison to developing countries is due to the differences in the death rates among developed and less developed countries, however, are not significant.

Table 50.2 gives the trends of fertility and mortality among developed and less developed countries of the world in the recent past. The table very clearly shows a faster decline of TFR in developed countries in relation to death rates.

Table 50.2 : Trends of Fertility and Mortality in the World, Developed Countries and Less Developed Countries (1950-95)

Period	*World*		*Developed Countries*		*Less Developed Countries*	
	TFR	*CDR*	*TFR*	*CDR*	*TFR*	*CDR*
1960-65	5.0	15.6	2.7	9.4	6.0	18.2
1965-70	4.9	13.4	2.4	9.4	6.0	15.0
1970-75	4.5	11.7	2.1	9.4	5.4	12.0
1975-80	3.9	11.0	1.9	9.5	4.7	11.5
1980-85	3.6	11.0	1.8	9.6	4.1	10.5
1985-90	3.4	9.6	1.8	9.6	3.8	9.6
1990-95	3.0	9.3	1.7	10.1	3.3	9.1

TFR = Total Fertility Rates, CDR, = Crude Death Rate (per 1000).
Source : World Population Prospects 1996 version UN 1995, p.20.

TFR and CDR in 1960-65 were 2.7 and 9.4 respectively in developed countries. However during 1990-95 CDR remained 10.1 very close to 1960-65 positions of 9.4, whereas the value of TFR declined to 1.7 from 2.7. On the other hand in less developed countries TFR and CDR both have registered a sharp decline from 1960-65 to 1990-95. The values of TFR and CDR in1960-65 were 6.0 and 18.2 in less developed countries, which went down to 3.3 and 9.1 respectively during 1990-95. The sharp decline of mortality in less developed countries is attributed mainly to the benefits of developments in the field of medical technology, which primarily took place in the West. However, the matching lower fertility could not be achieved in less developed countries due to lack of enough social and economic development. It has been pointed out that right from the beginning of the industrial revolution in Europe, cultural and other developmental factors kept fertility much below the biological maximum (Population Crisis Committee, 1985).

The influence of socio-economic development on demographic variables can be seen effectively by taking infant mortality rates (IMR) both for developed and less developed countries as shown in Table 50.3.

Table 50. 3 : Infant Mortality Rate (IMR) for World, Developed and Less Developed Countries 1960-65 and 1990-95

Year	*IMR*		
	World	*Developed Countries*	*Less Developed Countries*
1960-65	118	33	135
1965-70	102	26	115
1970-75	93	21	104
1975-80	87	18	98
1980-85	78	1	87
1985-90	69	13	76
1990-95	62	11	68

Source : World Population Prospects 1996 version UN 1998, p.7.

Table 50.3 very clearly shows a considerable fall in the IMR in developed countries during 1960-65 to 1990-95. It fell down to 11 from 33 during this period—something around 66 per cent. In less developed countries also the value of IMR declined but from 135 to 68 during the said period—something less than 50 per cent. As IMR is a very sensitive indicator of socio-economic development, the data given in Table 50.3 does indicate a higher socio-economic status of the developed countries. Thus we can say that as a result of development in medical technology in developed countries the benefits of reduction in mortality reach to developed countries as well as to less developed countries also. However, the benefits of social development do not reach to the developing countries in the same way. As a result there is no parallel fall in the fertility in the developing countries as will be the case in developed countries. Less developed countries will remain in second stage of demographic transition with rapid population growth.

In the light of the above theoretical and empirical relationship between population growth and development at a macro level for the world, the Indian situation can also be analysed more meaningful.

Population Growth and Development in India

Census of India in 2001 recorded the population of India about 1027.02 million as on 1st March, 2001 given it the share in the world population of about 16 per cent of from 15.2 per cent in 1981. Every sixth person in the world is an Indian. Since then, however, more than 100 ten million have been added to it.

Table 50.4 below gives the population of India for last 100 years according to present boundaries).

Table 50.4 : Population of India 1901-2001

Year	*Population (in millions)*	*Decadal Growth Rate*
1901	238.4	-
1911	252.1	5.7
1921	251.3	- 0.3
1931	279.0	11.0
1941	318.7	14.2
1951	361.1	13.3
1961	439.2	21.5
1971	548.2	24.8
1981	683.3	24.7
1991	843.9	23.5
2001	1027.02	21.6

Source : Census of India 2001 and earlier censuses.

Table 50.4 given above shows that during 1980 population of India increased by 161 million. The population of India in 2001 was around three times the population at the time of independence. The increase we add nearly the total population of Australia every year (Premi, 1991). The table also shows a consistent increase after 1921. The rate of population growth during 1921-31 was 11.0 per cent, which kept on increasing until 1971-81 when maximum of 24.7 per cent per decade. Only during 1981-91 and in 1991-2001 the decadal growth rate of population showed a marginal decline in relation to the previous decade and reached 23.5 per cent per decade and 21.6 respectively.

As along with developments in science and technology not only world population has increased, the means of subsistence have also been improved. In the following section, however, changes in some of the socio-economic variables having close relationship with population are discussed.

One of the direct consequence of rapid population growth is on the age structure of the population which will show a higher proportion of population in the younger ages. It will also affect the median age of the population. The population growth also indirectly influences the per capita availability of various resources. If proportional change in population and resources are same over a period time, the per capita situation will remain stable. However, if population growth exceeds the resources, the per capita availability will decrease and in the reverse situation per capita availability will increase.

From the view point of the age structure of the population the important age groups are 0-14 years (child population), 15-59 years (working age population) and 60+ years (elderly population). Percentage distribution of the population of India in these three age groups is given in Table 50.5 from1901 to 1991 as the age data for 2001 are still not available.

Table 50. 5 : Percentage distribution of population of India an deferent age groups 1901-91

Year	Age Groups			Median Age
	0-14	*15-59*	*60+*	
1901	38.60	56.35	5.05	21.3
1911	38.45	56.40	5.15	21.2
1921	39.20	55.55	5.25	21.1
1931	40.00	55.95	4.05	19.9
1941	38.25	56.85	4.95	21.0
1951	37.50	56.85	5.65	21.1
1961	41.00	53.36	5.64	20.5
1971	42.02	52.01	5.97	19.1
1981	39.54	53.98	6.48	20.5
1991	37.25	55.98	6.77	21.9

Source : In 1982 Population of India: Country Monograph series No. 10, New York.

Table 50.5 shows that share of 0-14 years has been very large. The table also shows a marginal increase in the 60+ years age group. The age group of 15-59, however, registered a marginal decline due to the increase in the group 60+.

As mentioned earlier changes in the age structure are the direct consequences of rapid population growth. Those populations which have higher proportion of population in the age group 0-14 years will have higher demographic dependency ratio. Population in the age group 0-14 is considered as children and our legislation does not allow them to work as regular worker. As the population in this age group will increase, the adult working age group population will decline proportionally. Dependency of adult population, however, is more of social nature.

Another direct relationship of rapid population growth is with the median age of population. The median age of population is also given in Table 50.5. Due to large concentration of population in the age group 0-14 years will increase, the adult working age group population will have greater burden to support them which in turn will slow down economic development and will bring more efficiency into the economy.

Another direct relationship of rapid population growth is with the median age of population. The median age of population is also given in Table 50.5. Due to large concentration of population in the age group 0-14 years, the median age is found to be quite low and due to this lowering of median age it is said that population of India is younger.

Another important variables which will have relationship with population growth, discussed here are literacy, employment, pattern of land utilization and production of agricultural goods. Their progress has been traced out during the post-independence period from 1951 to 1991 in the following sections.

Table 50.6 provides literacy rate and also total number of illiterates in India.

Table 50.6 shows an overall improvement in the proportions of literates to the total population which has increased from 16.6 per cent in 1951 to 40.7 per cent in 1991. Improvement in the literacy of female however is quite remarkable. It has registered an increase from 7.9 per cent in 1951 to 54.16 per cent in 2001— a phenomenal. The literacy of male during this period could also increase more than three times.

Table 50.6 : Literacy provides literacy rate and also total number of illiterates in India

Year	*Literacy rate in per cent*			*Illiterates in millions*		
	Total	*Male*	*Female*	*Total*	*Male*	*Female*
1951	16.6	24.9	7.9			
1961	28.3	30.9	15.3	333.9	148.5	135.4
1971	29.5	39.5	18.7	386.5	171.9	214.6
1981	26.7	39.5	16.8	424.2	182.6	241.8
1991	40.7	53.6	27.0	491.8	213.3	278.5
2001	65.38	75.85	54.16	450.3	191.3	269.0

Source : (i) M.K. Premi (1991) India's Population: Heading towards a billion, B.R. Publishing Corp. and Census of India 2001: Provisional Population Totals.
(ii) Statistical Abstracts of India 1997 (for 1951 values).

The story of illiteracy is also not different from that of literacy. Number of illiterates in India has also been increasing very consistently. Here again although female have shown higher improvement in the rates of literacy compared to male, their number of illiteracy has also increased and in 1991 it has exceeded the number of males. In 2001, however, for the first time there is a decline in the absolute figure of illiterates compared to the previous Census both for male and for female.

When we deal with the employment data we face the major conceptual problem of defining 'worker' which has undergone several changes in census definition. The National Sample Survey Organization, however, has given 8.35 per cent employment in 1971-73 which has gone down to 6.09 per cent in 1987-88 (Planning Commission, p.13). Sectoral division of workers as provided by census is relatively more comparable and is given in Table 50.7 for the period between 1961-91 only and the data required for 2001 is still awaited.

Table 50.7 : Sectoral Division of workers in percentages (1961-91)

Year	*Sectors*			
	Primary	*Secondary*	*Tertiary*	*Total*
1961	72.3	11.7	16.0	100.0
1971	72.6	10.7	16.7	100.0
1981	69.3	12.9	17.8	100.0
1991	67.4	12.1	20.5	100.0

Source : Census of India 1961, 1971, 1981 and 1991.

Primary Sector = Cultivators, agricultural labourers and live stock, forestry, fishing etc., and mining and quarrying.

Secondary Sector = Manufacturing in household and other than household and construction.

Tertiary Sector = Trade and commerce, Transport storage and communications and others services.

Table 50.7 shows a slow decline in the share of primary sector and a marginal increase in the secondary sector accompanied by substantial increase in the tertiary sector. Primary sector in the country has declined from 72.3 per cent in 1961 to 67.4 per cent in 1991. Tertiary sector, however, has shown an increase from 16.0 per cent to 20.5 per cent during the same period. Secondary sector which is the core of industrialization, however, has not shown significant increase in this period—although it has also improved marginally from 11.7 per cent to 12.1 per cent during 1971-91.

Effect of population is pronounced on land utilization, agricultural production and finally, on quality of life also. Percentage of net area shown, and area put to non-agricultural uses is given in Table 50.8 along with area sown more than once.

Table 50.8 : Changing pattern of land utilization in India 1950-51 to 1999-2000.

Year	*Total area (million hectares)*	*Percentage of net sown area*	*Per cent of area put to non-agri. uses*	*Net irrigated area*	*Percentage sown more than once*
1950-51	284.39	41.8	3.3	20.85	13.14
1960-61	298.46	44.6	5.0	24.66	19.57
1970-71	303.76	46.3	5.4	31.10	25.52
1980-81	304.15	46.0	6.4	38.72	34.82
1990-91	304.86	46.9	6.9	47.78	42.74
1999-2000 (P)	306.05	46.1	7.5	56.65	48.51

Source : Agricultural Statistics at a Glance-2003.

One of the important consequence of population of growth is the increase of net sown area as well as the area sown more than once. Table 50.8 shows as consistent increase in the percentage of net area along with a substantial increase in percentage of area sown more than once. The net sown area has increased from 13.14 per cent in 1950-51 to 48.51 in 1999-2000.

Another important consequence of population growth is on the land put to non-agricultural uses such as urban and industrial usage. The table also shows a consistent increase in area of land put to non-agricultural usage, which increased from 3.3 per cent to 7.5 during this period.

The consequence of population is more direct on food production and specially its per capita availability. Table 50.9 gives the total amount of production of cereals and pulses and per capita availability after indepencence.

Table 50.9 : Production of food grains and its per capita availability in India (2002-03)

Year	*Production (million Tonnes)*		*grams per capita per day*		*Total Food grain*
	Cereals	*Pulses*	*Cereals*	*Pulses*	
1950-51	50.22	8.41	334.	60.7	394.9
1960-61	82.02	12.70	399.7	69.0	463.7
1970-71	108.42	11.82	417.5	51.2	468.8
1980-81	129.59	10.63	417.3	37.5	454.8
1990-91	174.79	14.26	468.5	41.6	510.1
2001-02	212.03	11.08	386.2	30.0	416.2
2002-03*	187.57	11.31	454.3	34.9	491.2

* Provisional

Source : Agricultural Statistics at A Glance-2003.

The above table shows a phenomenal increase in the total production of food grain cereals which was 50-82 million tonnes in 1950-51 increased nearly four times in 1990-91 during this period to 212.03 mt. in 2001-02. However, it again discreased to 187.57 MT in 2002-03.

However part of this growth of cereals production offset population growth which is also about a little less than two and a half time. However, net availability of food grains per capita per day has been found to be continuously increasing but not substantially during the post indepencence period. It was 394.9 in 1950-51 which increased to 416.2 only in 2001-02.

Another important food grains are production of pulses which is increased only by one and a time during this period. Since the growth of population during 1951 to 2001-02 is about two and a half times, the per capita availability of pulses has gone down from 60.7 grams per day to 30.0 in 2001-02.

To summarize the above discussion we can say that although population is one of very important assets of any economy, rapid population growth of less developed countries may create a bottleneck for the development. Although in terms of relative growth of education and food supply etc. may have increased during the post independence period, it is very unlikely that this increase will be sustained in future also. A large proportion of growth in the economy is being offset by the growth in population. These are some sectors where per capita value has a decline along with the growth of population such as per capita available land and per capita availability of pulses etc.

Unsustainable burden of rapid population growth has its impact on the quality of life of the people also. Quality of life of an area hitherto was taken as synonymous to the per capita, "Gross Domestic Product" in many cases. However, in 1990 United Nations Development Programme (UNDP) introduced the concept of Human Development which should include many other aspects of human life also. Recognizing the difficulty in quantifying many of such variable, UNDP selected three most important variables which cover major part of the quality of life. These variables are : life expectancy. educational attainment and adjusted income (LTNDP, 1998). On the basis of these indicators, for which data is available for almost all the countries, UNDP worked out "Human Development Index" (HDI)

Table 50.10 : Life Expectancy (eo^o), Adult Literacy (ALR) her capita adjusted Income (PAI) and Human Development Index (HDI) of Selected Countries 1998

Countries	*(eo^o)*	*(ALR)*	*(PAI)*	*(HDI)*
Canada	79.1	99	21916	1
USA	76.4	99	26977	4
Luxembourg	76.1	99	34004	26
Kuwaït	75.4	78.9	23848	54
China	69.2	81.5	2935	106
Pakistan	62.8	37.8	22.9	138
India	61.6	52	1422	139
Sierra Leone	34.7	31.4	625	174

Source : Human Development Report, 1998.

and is being published each year in Human Development Report since 1990.

It is very important to note that position of India in terms of 'Human Development Index' is not very encouraging. Out of total 174 countries in 1998 report, India ranks 139th much below China which rank 106th. Pakistan in spite of substantial higher per capita real income is only one rank above India due to lower 'Adult Literacy Rate'. Its position in 138th. The ranks of 'Human Development Index' and 'its constituents of few selected countries are given below in Table 50.10.

As has been shown earlier, India has shown considerable progress in various fields of the economy. However, the rapid growth of population has been working as a big bottleneck and the country is not able to secure food security and good quality of life.

REFERENCES

Coale, A.J., and Edger Houver (1958), *Population growth and economic development in low income countires : A case study of India's prospects.* Princeton, N.J.: Princeton University Press, pp. 10-11.

Engelman, R. (1997), *Why Population matters.* Washington: Population Action Internal, p. 11.

Government of India, Min. of Agriculture (2003), *Agricultural Statistics at Glance.*

Kuznet, Simon (1966), *Modern economic growth.* New Haven ct: Yale University Press, p. 20.

Kirk, Dudley (1996), 'Demographic transition theory', *Population Studies,* Vol. 50, No. 3, pp. 361-87.

Mathus, Thomas Robert (1959), *Population : The First essay.* Am Abor Ann Arbor Paperback.

Mathus, Thomas (1960), *A Summary view of the principles of population in Three essays on Population* eds., Thomas Mathus, Julian Huxley and F. Osborn, New York, : Mentor Books, pp. 13-59.

Notestein, Frank (1945), Population : The Long view in T. Schutz (ed.) Food for the World. Chicago : (Normal Wait Harris Memorial Fund Lecture, 1985).

Planning Commission *Eighth Five Year Plan 1992-97,* Vol. 1, Govt. of India, p. 134.

Population Crisis Committee (1985), *Population growth and economic development,* Population Washington, D.C. : p. 1.

Premi, M.K., *India's Population : Heading towards a billion.* New Delhi : B.R. Publishing Corp.

UN (1953), *The Determinants and consequences of population trends.* Deptt. of Social Affairs Population Division, Population Studies No. 17, New York, pp. 56-61.

United Nations Development Programme (1998), *Human Development Report,* New York : Oxford University Press.

51

Ecology and Encounters

The Maratha Demands and its Impact on Land-use Pattern in the 18th Century Marwar (Rajasthan)

Vipul Singh

Introduction

The ecology of Rajasthan has tremendously influenced the social, political and economic life of the people of this region. George Thomas was the first person to have called the geographical region in the western part of India as 'Rajputana' in 1800 A.D.[1] Col. James Tod gave it the name 'Rajasthan' which included 21 princely states—Udaipur, Dungarpur, Banswara, Pratapgarh, Shahpura, Karauli, Jaisalmer, Bundi, Kota, Sirohi, Jaipur, Atwar, Jodhpur, Bikaner, Kishangarh, Danta, Jhalawad, Bharatpur, Dholpur, Palanpur, Tonk.[2] Rajasthan before the 17th century was not known by any single common name for long. The Mughals knew it as Suba Ajmer.[3] The first mention of Rajasthan as compact land of the princes with territorial divisions, plains and mountains occur in the Inscription of 1708 AD.[4]

The inscription has these lines :

Desh dharm kshetra sagar sapavitra kshetra tanmadhye
Meru shikhar saraj vijay Rajasthan sanrippanivasah

The British gave the name Rajputana for the administrative purposes to a huge area which included eighteen native states

(Jaisalmer, Jodhpur, Bikaner, Alwar, Jaipur, Bhartpur, Dholpur, Karauli, Bundi, Kotah, Jhalawar, Pratapgarh, Banswara, Dungarpur, Udaipur, Sirohi, Kishangarh and Tonk), two Chiefships of Shahpura and Lawa[5] and the British district of Ajmer-Merwara. The word Rajputana was an adopted form from Rajputan, the plural of Rajputs in Persian.[6] Thus, here', by 18th century Rajasthan we mean administrative unit, which included Suba Ajmer of the Mughals as well as independent and semi-independent principalities. Deryck O. Lodrick says that this treatment of Rajasthan as an administrative region is an intellectual concept with some purpose of thought, but since here in this paper we are trying to trace the contribution of the state towards land degradation we cannot do away with it.[7]

Although Deryck O. Lodrick has given a somewhat distinct explanation for the creation of Rajasthan as a region, he says that the desert, the dress of the people, and the Rajput clans have not only helped in creating an image of Rajasthan, they are also elements of a specific environmental setting, cultural tradition and social structure that have all played a significant role in shaping the character of modern Rajasthan.

The geographical location of Rajasthan is between $23^{o}\ 3'$ to $30^{o}\ 12'$ in the north and $67^{o}\ 30'$ to $78^{o}\ 18'$ in the east. Roughly, the shape of Rajasthan is like irregular rhombus. The territory is integrated diagonally into two parts by the Aravali range running from north-east to south-west. However, there is a little break in the Aravali range where the city of Ajmer is situated. The upper part or the western part (we may call it) occupies about two-third of the area of Rajasthan and includes the states of Jodhpur, Bikaner and Jaisalmer. This western part, which is an assemblage of many states, is comparatively well drained and fertile.[8] In the eastern side of Jodhpur lies the state of the Kachwahas with its independent offshoots, the Shekhawati state and Alwar. In the southern part of Rajasthan lies the historic land of Mewar with the vassal states of Dungarpur, Banswara and Pratapgarh. In the south-east along side Mewar is Harauti, which includes the states of Kota and Bundi. G.N. Sharma says that Rajasthan during the 18th century comprised independent and semi-independent principalities like Jodhpur, Jaisalmer, Bikaner, Ajmer, Nagor, Sirohi, Udaipur, Dungarpur, Jaipur, Alwar, Bharatpur and Bundi, which were known as Maru, Mada, Jangaldesh, Ahicchatrapur, Arbud, Medpat, Vagad,

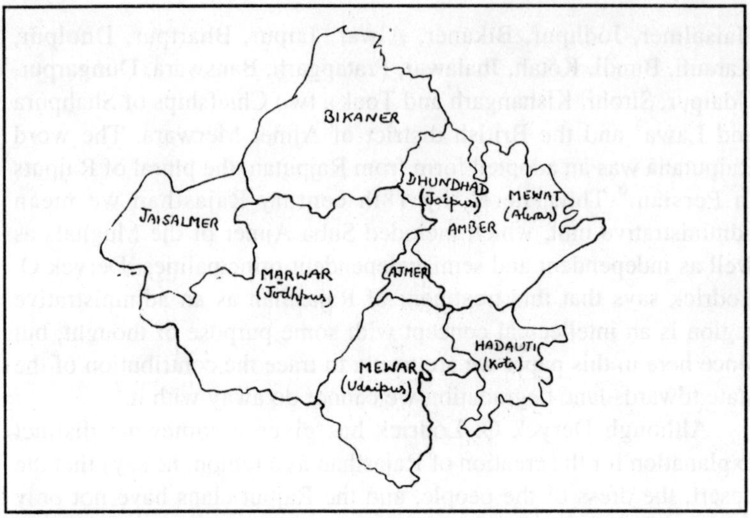

Fig. 51.1 : 18th Century Rajasthan.

Dhundhar, Mewat and Hadoti respectively during the early medieval period.[9] (Fig. 51.1) He has been able to dig out the earlier names of other areas as well. He says that the tract on the Mahi river around Pratapgarh was called Kanthal. It was so called because it formed the border *(Kanth)* between Mewar on the north, Vagad on the west, and Malwa on the east and the south. The land between Devaliya (Pratapgarh) and Banswara consisted of fifty-six groups of villages and, therefore, that tract of land was called Chappan. The land between Dungarpur and Banswara was called Mewal. The area between Devaliya and Mewal was called Mundal. The area between the Jarga and Rahagra mountains was called Deseharo (land covered with green grass). Another name for the area around Udaipur was Girwa, land surrounded by *giri* or mountains.

Physically, Rajasthan has four divisions— (a) the western sandy plains, (b) the Aravali range and hilly region, (c) the eastern plains and (d) the south-eastern plateau. The Thar Desert makes up the western plains of Rajasthan, which spreads from the western slopes of the Aravali up to Sind. These areas have very little vegetation consisting mainly of thorny bushes and are sparsely populated. Climatological studies have found abnormal rainfall in this region as common. Monsoon failures occur during 40 per cent of year, and normal rainfall during monsoon occurs only during 26 per cent of the weeks when it is expected.[10] Droughts leading to agricultural failures and famines have occurred throughout the historical period of settlement.[11] The annual rainfall averages from 250 mm to 350 mm and so, crop cultivation in this part is carried out for subsistence production of food and fodder, along with livestock rearing. The water tables in Western Rajasthan are low and range from 60 to more than 100 metres and ground water tends to be saline, brackish, and unfit for drinking or irrigation.[12] The soil of this region is light-textured, gray-brown desert soil, which are alkaline and saline and thus is not suitable for agricultural production. The soil type is sandy in majority of the region except near water sources where we find light alluvium type soil. Such soil type is highly susceptible to wind erosion. The wind erosion in many areas tends to destroy the newly germinated crops. The little vegetation that is possible in this area throughout the year is that of *khejri* and we get reference to *khejri in* many historical documents. Luni is the only major river of semi-permanent nature in this area and flows during the rainy season only. The Aravali range and hilly region extends from Sirohi to Khetri and stretches up to Delhi. Geologically it is one of the most ancient hill ranges of the world.[13] It covers Sirohi, Udaipur, Banswara and Dungarpur districts. The Aravali hill system has greatly influenced the climate of Rajasthan and has also been a check against the process of desertification from the western side of Rajasthan to the eastern. But the hill only acted as a constraint and checked the spread of desert till it remained green and well forested.[14]

The eastern plains extend from the east of Aravali range to the plains of the rivers Ganga and Yamuna. Its eastern boundary is marked by the eastern edge of the Aravali up to the north of Udaipur. The Vindhyan plateau marks the south-eastern limit of the plains. The area is comparatively well off in agricultural production. In fact the Maratha penetration has mainly taken place in these states falling east of the Aravali range, i.e. Mewar, Jaipur, Kota and Bundi and only one state west of the Aravali, that of Jodhpur. It is basically because of this geographical setting and location of Aravali range that states like Bikaner remained safe. Its physical features came to its rescue.[15] In view of the obstruction offered by the physical features and the climate of the State of Bikaner, outsiders might never have felt tempted to come to the area and the history of Bikaner might have passed mostly as record of uneventful events confined usually to issues of local importance alone. The Rathore principality of Bikaner offered so inhospitable a region that only a few outsiders dared penetrate the sand dunes and disturb the peace of the area. The Bikaner State remained immune from attacks from outside. Neither the Maratha nor the Pindaris ever disturbed the peace.This offered a very great security to the area from external attacks and allowed the rulers of Bikaner to shape the destiny of their principality independently. It meant an autonomy well suited for the growth of an indigenous culture and undisturbed harmony. The sense of security provided to the people residing in the area had primarily been responsible for the prosperity of *Seths*, *Sahukars* and other mercantile communities though in every other respect the waste lands of Bikaner offered practically no opportunities for any prosperity to flourish. The *Seth* and *Sahukars* of Bikaner who conducted trade on all India basis and resided mostly outside the state in distant places like Calcutta and Bombay kept their wealth hoarded in Bikaner primarily because of the security provided by the physical features. The security provided by nature induced literary people to shift their manuscripts to this area and that explains the presence of thousands of manuscripts even today in the Bikaner state Archives.

Rajasthan was economically underdeveloped during the late medieval period as compared to the neighboring Malwa, Gujarat, Delhi and Agra because of ecological limitations. Nevertheless, its geographical situation made it strategically important to any ruling authority. During the Mughal period its close proximity to the heartland of the Mughal Empire had made the control over it crucial for the Mughal Central Government. The economic signification of the region lays in the trade routes connected the upper Gangetic plain with the seaports of Gujarat passed through its territories. The close integration of Eastern Rajasthan with the Mughal state structure and economic processes at work during the 17th century brought about a marked growth in its economy. But, the decline of the Mughal Empire in the century majority affected the socio-economic structure of Rajasthan. The political map of Rajasthan began to change rapidly with the ever-increasing weakness of the Mughal state structure after the death of Aurangzeb. Each of the various Khalisa *parganas* in the Ajmer

Suba was either being granted in *Jagir* or was forcibly being occupied and acquired on lease by the powerful Rajput rulers of the states like Mewar, Marwar, and Amber, etc.

Marwar is a corrupt form of Maroowar, meaning the region of death[16] and it came to be known as Jodhpur only after the city of Jodhpur was founded in 1459 A.D. Marwar is placed in the western part of Rajasthan and lies between the latitude 24° 36′-27° 40′ and longitude 20°4′-75° 23′. It is the largest Rajput state of Rajasthan and bounded on the north by Bikaner and Shekhawati, on the north-west by Jaisalmer, on the east by Jaipur, Kishangarh and Ajmer, and on the south by Udaipur and Sirohi. Physical diversity is one of the many features of Marwar with a basic contrast in geographical set-up of the western and the south-eastern parts. The western portion is purely a desert whereas the other half is relatively more fertile. R.C. Sharma uses the term 'little desert' for the later part.[17] The pattern of settlement in area such as this largely depended on the character of water supply. Since the eastern and the south-eastern portions has higher water table, the settlement tended to be denser than the remaining parts where it was much lower.[18]

Discussion

Traditionally the Rathors have been the rulers of Jodhpur state. They claim their descent from Rama of Ayodhya and the original name of their clan was Rashtrakuta, which later on crystallised into the word Rathor.[19] The Rashtrakutas ruled over the Deccan area during the 9th and 10th centuries, but were ousted by the Chalukyas in 973 A.D. and so they moved to Kannauj. Here they founded a new dynasty of Gahadwala (Gaharwar). The last of the Gahadwala rulers, Jai Chand, was defeated in 1194 A.D. by Mohammed Ghori. Siaji and Sait Ram were the two grandsons of Jai Chand and they decided to give their services to the chief of Kolumund near Bikaner. This was the period when the small principalities of Western Rajasthan were constantly fighting with each other. Gradually, the Rathor Rajputs won the confidence of the rulers of Pali by giving their services, as they were quite successful in maintaining order (*Zabta*) in Pali territories. The disturbances posed by the Mer tribesmen were also reduced. Rathor *sardars* were given land in the neighbouring areas and with the matter of time they gained so much strength that they captured many surrounding areas. By the time of Rao Chand (C.1383 to 1432 A.D.), the Rathore chief, they ruled over Pali, Khed, Bhadrajeen, Kodna, Mahawa (Mallani), Barmer and some areas of Bikaner.[20] Rao Chand had made Mandor the Capital in 1381 A.D., but in the time of Rao Jodha the seat of government was shifted to Jodhpur 1459 A.D. Thus, the period of 13th and 14th centuries that started with mutual agreement between the Rathor Princes and the local chiefs of Pali got transformed into Rathor dominance and sovereignty at the end. In the process of their expansion the Rathors followed the practice of *bhai-bant*.[21] Under it, the sons and brothers of the ruler used to occupy various territories won over by them and these territories were termed as *thikanas.* These *thikanedars* had a kind of political and normal obligation to provide military service to the state. But obviously one cannot expect very strong and honest service by the *thikana* holders to the state throughout the medieval period. To a large extent it depended on the strength and competency of the individual ruler and how far he was able to keep his clan members in good humour. In normal circumstances, the system of *bhai-bant* might form a true political entity, able to acquire and maintain some sort of rule within the state and even beyond the state borders. But this process was in no way automatic and so simple to achieve for a ruler. The ruler to be successful would never grant any members of his clan a domain large enough to challenge his authority. At the same time he needed these clan members because without them he would have no chance of withstanding the claims of neighbouring states. The Rathore *sardars* also allowed the Rajputs of other clans and the Jats to settle in their *thikanas.*[22] The logic behind insisting these new groups in the conquered areas was probably to increase the numbers of people loyal to them and to extend the area under cultivation. The political set-up of the Marwar state was such that only a strong and diplomatic ruler could keep the *thikanedars* united and obidient to the state. The Rathor *sardar* could declare themselves independent under a weak ruler and in fact such a situation arose in the state after the death of Rao Jodha in 1484 A.D.[23] It is only when Rao Maldeo became the ruler in 1532 A.D. that the political environment of the state changed considerably as he was very strong. He added forty *parganas* to his hereditary dominion of Jodhpur which included Ajmer, Barmer, Didwana, Bikaner, Jodhpur, Sanchor, Jalor,

Jaisalmer, Sojhat, and Merta.[24] He even conquered the neighbouring Bikaner state in 1541 A.D. Although Rao Maldeo's power was reduced after his defeat at the hands of Sher Shah and he lost Sambhar, Didwana, Lalsot, Chetsu, etc., but regained these areas after Sher Shah's death and occupied Barmer, Kotada and some parts of Jaisalmer territory. Rao Maldeo had so much increased his power that he started the practice of assigning *pattas*, i.e. land revenue grants to his Rathor kingsmen who served in his army. There was a difference in assigning *thikanas* and giving *patta* grant to his supporters. Under the *thikanas*, the Rathor sardars had some kind of autonomy for internal administration, whereas the *pattas* were assigned purely in lieu of their service. The *pattadar* (*patta* holder) was responsible for the collection of revenue and civil and judicial administration. If we look at the functioning of the *pattadari* system it looks very much similar to the *Iqtadari system* of the Delhi Sultanate. But there is a major difference. The nature of the holdings of Rathor *pattadari* was more permanent in nature because of their kingship ties and clan-solidarity. But again the *patta* system could be successful only when the ruler was powerful enough to keep the *pattadars* under control. Rao Maldeo was able to handle both the *thikanedars* and the *pattadars* very successfully. But Rao Maldeo's death in 1562 A.D. led to political instability in Marwar as his sons started fighting with each other for *gaddi* of Jodhpur. This was the period when the Mughal Empire under Akbar was trying to take control of the neighbouring territories. Akbar had earlier brought Jaitaran and Merta in Rajasthan under his rule[25] and now he got the opportunity to intervene in Jodhpur state as well. Among Rao Maldeo's sons—Rao Ram, Udai Singh and Ravimal, Rao Ram was the first to approach Akbar for support. Akbar recognised the hereditary right of Rao Ram and sent Hussain Quli Khan to march against Jodhpur along with Rao Ram. They first captured Sojat and then seized Jodhpur. Gradually, Akbar brought whole Jodhpur territory under the *Khalisa* and gave *Jagir* to various Rathore *Sardars*. He even discontinued the hereditary claims of a ruling family over the *gaddi* of Jodhpur. He made it clear before the Rathore *sardars* that as a sovereign ruler it was his prerogative to choose the person for *gaddi.* Thus, the internal rivalry and conflict among the Rathor *sardars* had led to their downfall. However, many of the sardars joined the imperial service at the Mughal court during

Akbar's period and were given *Mansabs* and *Jagirs*. But not all these *jagirs* were *watan Jagirs* or *des-ra-pargana* (ancestral territory).[26] Another kind of *Jagir* came into play called *tankhwa jagir*. When Marwar accepted the suzerainty in 1563 the then ruler had paid 4 lakh *Pirojiya* as *peshkash*.[27] And now when they were bound to obey the order of the Emperor as any other noble, the estimated revenue of the area called *jama* was adjusted against the salary of the ruler's *mansab*.[28] As *watan jagir* could not be transferred and was permanent in nature, the *Rathors* were given *tankhwa jagir* in lieu of the balance of their salary.

When Udai Singh ascended the throne in 1583, Akbar gave him a part of *pargana* Jodhpur. Gradually he strengthened his position through matrimonial alliance with the Mughals. He married his daughter to Prince Salim.[30] Munhta Nainsi says that at the time of his accession Udai Singh was given the *mansab* rank of 1000 *zat* and 800 *sawar* and at the time of his death he reached the rank of 1500 *zat* and 1500 *sawar*.[31] The successors of Udai Singh were able to get further enhancement in their *Jagir*. Raja Suraj Singh got the *mansab* of 5000 *zat* and 5000 *sawar*. And when Gaj Singh became the Raja of Marwar he further got the *parganas* of Phalodi, Jalor, Sanchor, Merta and Nagaur.[32]

Under the rulership of Jaswant Singh (he got the *gaddi* of Jodhpur in 1638 A.D.) the Rathors began to regain their firmer hold over Marwar. Under the rulership of Shah Jahan, he gradually got the *mansab* increased for himself through his services to the Mughal state. At the time of his succession he was granted the *mansab* of 4000 *zat*/ 4000 *sawar*, which was raised to 7000/7000 of which 5000 were *du-aspa, sih-aspa*, by 1657 A.D. So within 20 years time he was able to get all the *parganas* of Marwar in *jagir*.[33]

Jaswant Singh even got the *jagirs* outside Marwar. The following Figures in Table 50.1 give the clear picture of the enhanced power of Jaswant Singh.

However, the problem arose when Jaswant Singh died in 1678 A.D.[34] without any heir to the throne and the succession of the *gaddi* of Jodhpur became a bone of contention. Although the Mughal emperor Aurangzeb immediately issued a *farman* stating that the *desh* Jodhpur would be given to the son of the Maharaja, who has met an untimely death, if born to one of the *Ranis*. But by the time a son Ajit Singh was born,[35] the politics within the court had changed the

Table 51.1 : Jagir outside Marwar

Year	*Mansab*	*Jagir in Marwar*	*Jagir Outside Marwar*
1638	4000/ 4000 *Zat/Sawar*	3,99,80,000 *dams*	Nil
1657	7000/ 7000 *Zat/Sawar* (5000 *du-aspasih-asp*)	6,32,25,000 *dams*	3,44,95,000 *dams*
1674	7000/7000 *Zat/Sawar* (5000 *du-aspasih-asp*)	7,17,52,731 *dams*	61,74,522 *dams*

Source : Based on the data available in *Marwar- ra- Pargana- ri- Vigat*, Vol. II and *Jodhpur Hukrunat- ri- Bahi*, Meerut, 1976.

situation. One of the Rathor sardars, Indra Singh, wanted to get the *jagir* of Jodhpur in his favour and in fact Aurangzeb probably moved away from his earlier stand of appointing Ajit Singh as the successor. Many Rathor *sardars* were won over to Indra Singh's side by grant of *pattas*. But Durgadas Rathor and few other *sardars* refused to accept Indra Singh as the ruler and were adamant on making Ajit Singh as the ruler. The presence of Durgadas and other Rathor *sardars*, and the fact that the young Maharaja (Ajit Singh) was in Marwar provided a favourable atmosphere for the unity of the Marwar *sardars*.[36] Further, the declaration of Rana Raj Singh of Mewar to support the claim of Ajit Singh drastically ignited the political environment of Mewar. Mewar ruler had his own political compulsions to support the cause. He might have feared that grant of *gaddi* of Marwar to Indra Singh could lead to the conversion of the territory in *Khalisa* and the next target of the Mughals could be the neighbouring Mewar territory. Rana Raj Singh gave protective shelter to Ajit Singh and even sent 5000 *sawars* to help the Rathors.[37] But the death of Rana Raj Singh in 1680 changed the scene once again. His successor Rana Jai Singh was of the belief that Mewar's support to the Rathors was needlessly souring enemity with the Mughals. So he asked the Rathor *sardars* to vacate Mewar.

In the mean time Prince Akbar declared himself as the emperor of the Mughal empire revolting against Aurangzeb. Marwar *sardars* under the leadership of Durgadas offered their support to him.[38] But the combined forces of Akbar and the Rathors were defeated by the

Mughal army in 1681. Durgadas had to go to the Deccan along with Akbar. After Durgadas' departure it was the Sonag Rathor who became the leader of the Rathor *sardars* and they started plundering the imperial posts and realized land revenue from the cultivators in Marwar. The Mughal emperor probably proposed to grant Jodhpur to Ajit Singh and a *mansab* of 7000/7000, but it was just to deceive the Rathors—जोधपुर वगैर मुलक ने सात हजारी मनसब, जसवंत सिंहजी रा दस्तूर अजीत सिंहजी नु देसा.[39] The death of Sonag Rathor again pushed the Rathors one step backward. They were once again in a look out for a leader in the absence of Durgadas. They now got the leader in Udai Singh Rathor and then the Rathors decided to bring Maharaja Ajit Singh and make public appearance, who was till now in concealment.[40] Ajit Singh along with Udai Singh Rathor and other *sardars* went to Ahwa, Began, Rajpur, Bilada, Balunda, Riyan, Asop, Lavera, Hinwsar, Kolu and other places, and accompanied by the *sardars* of these places proceeded to Pokaran.[41] The Rathors not only attacked various Mughal posts but also collected revenue from different areas. By now Durgadas too had come back from the Deccan to Marwar.[42]

Mughal Decline and Economic Pressure

The first half of the 18th century saw the disintegration of the Mughal empire. Irfan Habib, Athar Ali and Satish Chandra have explained different structural reasons for it. Although all of them have their own independent view on the issue, but broadly there is unanimity among them.[43] The decline of the Mughal power gave opportunity to different regional centres to enhance their own authority at the expense of the Mughal power. The Rajputs in Rajasthan and the Marathas in Maharashtra were no exception to this. The Rajput states in Rajasthan resumed their earlier process of conquest and expansion which had actually stopped during the second half of the 16th and 17th century under strong centralised Mughal rulership. Sawai Jai Singh of Amber cherished his ambition of establishing his hegemony from Sambhar lake in the west to Agra and Mathura in the east and the Narmada in the south.[44] He took control of *pargana* Bhangarh, Malarna, Amarsar, Umara and Barwar. In 1729 he annexed Rampura district from Udaipur and attempted to bring Bundi under his vassalage by setting up his own nominee on its throne.[45] Similarly, Maharaja Ajit Singh of

Jodhpur brought many territories under his control. This had led to a rivalry between Jai Singh and Ajit Singh. Taking advantage of the declining power of the Mughals, the Marathas also started their gradual movement and activity in the north and captured Gujarat and Malwa. They frequently raided Rajasthan and even reached up to Delhi. In Rajasthan they collected *chauth* and *Khandani* from the Rajput rajas. With the increasing power of the Marathas and the declining authority of the Rajputs, mainly due to internal feuds, these raids for revenue collection became more frequent and adversely affected the economy of the states like Marwar and Mewar. The effect could have been disastrous in an arid zone like Marwar about which we shall discuss in this paper ahead.

The financial status of the state was already depressing because of the geographical factor as well as the decrease in the extra revenue coming to the state in the form of *tankhwa Jagir*. Dilbagh Singh has shown an actual contraction of cultivation in eastern Rajasthan mainly because of the Maratha raids in the latter half of the 18th century.[46] S.P. Gupta further elabortes on this issue and says that another major factor for decline that operated now was surely the loss of Rajasthan as the favoured province of an 'all-India Empire'. With the latter's disappearance, the external revenues ceased to flow in, and internal pressure for revenue correspondingly increased, with its own self-defeating consequences.[47] It is important to note here that the situation of Rajasthan was different from that of other Mughal territories, viz. Bengal, Oudh, etc., in the 18th century and so when we talk about the political and socio-economic decline of Rajasthan, we must consider the continuation to some extent of the Mughal administrative system or the administrative system of different Rajput states. In the case of other provinces like Bengal the interference of the East India Company started in the second half of the 18th century and so the region went through a phase of decline.[48] But so far as Rajasthan is concerned, the Rajput ruling houses continued to rule during this period and interference of the East India Company could be observed only towards the beginning of the 19th century.[49]

The declining political and socio-economic condition created a disastrous situation for the agrarian system and the land-use in Rajasthan. The production, distribution and exchange relations between the landowners and the land users became all the more exploitative. The land use pattern it encouraged was not related to the land-use capabilities and hence was over exploitative. The net result was that the need for preventing very intensive use of land was completely overlooked and the conservation needs of the resource base never figure in the agrarian system and measures of the Rajput rulers. These phenomena were further aggravated by large scale migration of cultivators, especially the *raiyatis* to the more prosperous areas of Malwa and Harouti, leading to an increase in number of villages classified as *Ujar*, i.e. desolated.[50] It was a normal tendency of the cultivators to leave fallow the poorer quality land in such a situation.[51] The situation was all the more alarming in the region west of the Aravali ranges. The climate in this region was quite arid and there were frequent famines. Further, 18th century in Rajasthan was a period of state formation. Many *thinkanas*[52] came up and Rajasthan was divided into many small princely states. This may also have been a major factor in the transformation of agriculture or forestland into dry deserts.[53] We will have to see as to how far was agriculture linked to state formation and whether this actually affected the ecological balance.

After the death of Ajit Singh, the political situation in Marwar became very fragile. His successor Abhay Singh had to face opposition from within the Rathor nobility and so he did not feel very strong as a ruler, even though Muhammad Shah had officially granted him the *gaddi* of Jodhpur and a *mansab* of 7000. The decline of the Mughal power and the increasing attacks and raids by the Marathas created a difficult situation before Abhay Singh financially as well. Now the *jagirs* granted to him outside Marwar yielded only meagre revenue and, whatever he was able to extract, that too was not regular. The Maratha raids into Rajasthan and the internal bickering that ensued after Ajit Singh's death made it compulsory for Abhay Singh to be constantly present in Marwar. His absence could have led to his overthrow or greater political manoeuvering within Marwar. It is important to remember that the policy of Ajit Singh after 1707 A.D. had actually created a conflicting situation within Marwar. In order to get rid of the Champawat Rathors, who had become very powerful, he promoted the Bhandaries. Although this was actually a defensive measure to create his own political basis and to avoid conflict, ultimately it proved fatal when Abhay Singh came to rule and it led to civil war. Land rights at that time were the most influencing factor in all the major upheavals in Marwar. Land rights in terms of *patta* were both the social and economic basis of any ruler. Ajit Singh had disturbed this socio-economic basis slightly by transferring, *pattas* to different new *sardars*. G.D. Shamra believes that this was the primary cause of the rebellion after Ajit Singh's death.[54] The frustrated dislodged *pattardars* were not having any holdings to meet their basic requirement and so they revolted when a new ruler came. Because of these internal revolts Maharaja Abhay Singh could not render any service to the Mughal emperor outside Marwar. This must have affected the economy of the state since *tankhwa jagir* outside Marwar was some sort of bonus for the state. These changes were observed more after Abhay Singh's death under Maharaja Vijay Singh during whose reign the Maratha raids had become more frequent. Abhay Singh had introduced a new practice in the land rights. He started collecting lump sum amounts from the *pattadars* to pay a huge amount to the Marathas and to maintain his own internal administrative expenses.[55] Before we move on to discuss the new changes introduced during the phase of Maratha incursion in Rajasthan, let us first discuss the nature and types of land rights and the sources of revenue of the state. In this process we shall also look at the nature of relationship between land and society.

The Rathors from the beginning of their rule in Marwar had believed in the concept of *bhai-bant,* which meant that all the Rathors (blood-relations) are equal. The Rathor ruler was first among the equals. But gradually the ruler of Marwar emerged powerful and Maldeo began asserting himself among his clansmen. He refused to accept the concept of *bhai-bant* and started treating the Rathor sardars as *bhai-bant chakar*, implying the subordinate position of the assignees to whom different territories were given.[56] The system of *bhai-bant was* now replaced by the *pattadari* system or revenue assignments. This was a totally new relationship between land and society and the territories were given to the Rathor clansmen through *pattas*. The *patta* holders called *pattadar* were obliged to render military service within and outside Marwar. They were free to manage their respective territories.

During the 17th century Marwar came under Mughal rule. However, the Mughal emperor gave the Rathor rulers some kind of autonomy to rule their territory.[57] They were given *Watan jagir*.[58]

The Rathors were also assigned additional ordinary *jagir* outside their *watan*. The *watan jagirdars* were given autonomy to administer their territory and even sub-assigned it to others. These Marwar rulers were allowed to transfer their rights to land revenue and other taxes to their subjects and their clansmen through the grants of *patta*.[59] The *patta* was assigned to different Rajput *sardars* against their salary by calculating the income from the land assigned. This calculation was often in terms of *rekh* and not *jama* of the Mughal system of administration. During the 17th and the 18th centuries the connotation of the term *rekh* was more or less-identical with the term *jama* or *jamadami*, but the mechanism of assessing in *rekh* was quite different from *jama*. In the Mughal Empire, before assigning the *Jagir* to a *mansabdar*, an estimated net income from the land revenue of the territory was prepared and the *mansabdar* was given the *jagir* equivalent to his salary.[60] This estimate was known as *jama*. Another term that was frequently used in this connection was *hasil*, which was the actual realization, made by the *jagirdar* from the assigned territory. However, the Rajput states in Rajasthan had the unique system of fresh assessment of estimated revenue and this was done as *rekh*.[61] Since in Marwar a *jagirdar* (both *watan jagirdar* and normal *jagirdar*) would subassign a part of his *jagir*, he needed to have some estimate of revenue on the basis of which he could subassign. It were these figures of assessment of different villages on the basis of which *patta* was issued that were known as *rekh*. It is nowhere mentioned in the primary accounts like *Bahis* or *Marwar-ra-Pargana-ri-vigat* how the figure of *rekh* was worked out. But apparently some estimation of revenue realization minus the cost of collection must have served for its basis.[62] It was quite natural that the *rekh* figure would be higher than *jama* figure, because a *jagirdar* assumed that the land revenue on the basis of which he was subassigning should not be less than the *jama* at which he had received the assignment. Munhata Nainsi has given the statistical details of *jama*, *rekh* and *hasil* for the whole of Marwar, which would give a clearer picture of the three terms associated with land revenue. *Jama* signified the actual capacity of the peasantry to pay, while *rekh* figures were the maximum assessed revenue of the area.[63]

Table 51.2 : Jama, Rekh and Hasil of Marwar

Year	*A* *Jama*	*B* *Rekh*	*C* *Hasil*	*% of C to A*	*% of B to A*	*%of C to B*
1660	1580625	2586941	1570575	99%	163.6%	60.7%

Source : *Marwar-ra-Pargana-ri-Vigat.*

The *pattadar* was authorized to collect other types of taxes as well, other than land revenue *(bhog), like ghasmari* (grazing tax), *jihat* or *Kharach bhog* (expenditure of collecting the land revenue taken as tax from cultivators), *kayalt* and *tolvantai* (weighing tax) and *mal* (some kind of tax other than land revenue).[64] There is another related term, called *lag-bag*, which has been frequently used in *Marwar-ra- Pargana-ri-Vigat*. The term *lag-bag* implied miscellaneous cesses including the taxes to meet the expenses incurred in the collection of the land revenue and the land revenue and the tax levied on pasture land.[65] It may also be noted that peasants had to pay many other kind of agricultural and non-agricultural taxes. In *Sanad Parwana Bahi* of late 18th century we get reference to grazing taxes like *pan-charai, patra-charai, ghascharai, uoont-charai, bhains-barar, bakra-bab,* etc.,[66] Col. James Tod mentions about a tax to be paid by the cultivators on the number of ploughs used by them and it was Rs. 5 per plough in Marwar.[67] During the second half of the 18th century, probably after the pressure of heavy Maratha demands, the state started collecting special tax for the maintenance of revenue records called *likhvani.*[68] *Sanad-Panwana-Bahi* also mentions that the government appointed a headman in each of the villages known as *choudhary* to assist in the assessment and collection of land revenue. The cultivators were supposed to pay a tax called *choudhary-bab* to be given as renumeration to the *choudhary.*[69] Further, the state made discrimination in collecting revenue from different castes of cultivators. Those who belonged to the higher caste enjoyed certain concessions in revenue as compared to the ordinary peasants. The following figures of revenue burden on the peasantry of different castes worked out on the basis of information provided in *Marwar-ra-Pargana-ri-Vigat*, give a good look on the differential scale of revenue assessment adopted by the state.

Table 51.3 : Revenue Burden on Caste Peasantry

	Rajputs	*Banias*	*Ordinary*
Revenue realized as % of produce	25 %	33 to 37%	35 to 40%

Maratha Demands

The Maratha interference had already started in the second quarter of the 18th century in various states of Rajasthan. But this had not reached to such a proportion so as to affect the economy of these states directly during this period. There are stray references of the Marathas getting money in return for their help to either of the aspirants to the throne of the Rajput states, but these were not regular payments and did not force the rulers of Marwar to introduce any new measure and policy to increase the revenue of the state to adjust the extra expenditure incurred on account of their payments to (demands of) the Marathas till the first half of the 18th century. It is only in the second half of the 18th century that we start getting regular account of payments to the Marathas in terms of *mamlat, Nazrana, Barat, Lashkar Kharch, Fauz Kharch, Darbar Kharch. Khasa Sawari*, etc. Dilbagh Singh has presented before us the case of eastern Rajasthan suggesting that the political weakness of the Rajputs paved the way for the entry of the Marathas into Rajasthan, and to dominate the area in course of time. He is of the opinion that the Maratha incursion together with frequent appearance of famines had serious effects on the agrarian economy of the region during A.D. 1750 and 1800.[70] The Marathas squeezed the already shrinking revenue resources of the state by their constant demand for huge monetary sums by way of war indemnity and annual obligation of *peshkash.*[71] The case of Marwar was not very different. Here too the Marathas squeezed the revenue resources by their frequent demands for money. But the overall pressure on the economy was relatively greater in comparison to eastern Rajasthan owing to the geographical location of the state. The state fell in semi-arid and arid zone of Western Rajasthan and the agricultural land available here was very limited. These agricultural lands had a limited production capacity owing to the total dependence on rain and non-introduction of new agricultural technology during these years.

In the second quarter of the 18th century the money that was being given time-to-time to the Marathas was neither regular nor was it based on any agreement or treaty. But the situation changed in the later half of the 18th century. In February 1756 a treaty was signed between Maharaja Vijay Singh of Marwar and Jankoji (Maratha *Sardar*) and it was agreed upon that the Marwar rulers would pay annual tax called *mamlat* to the Marathas.[72] The then ruler of Marwar, Maharaja Vijay Singh promised the Marathas that he would pay Rupees one lakh fifty thousand every year and this included *nazrana* as well.[73] Apart from these regular annual payments, the Marathas demanded heavy amount time to time depending on the changing political scenario of the Jodhpur state. The burden of these payments used to be so high that the Marwar ruler was never able to satisfy the Maratha demands completely and there always remained a part of the promised amount as due leading to greater burden on the economy of the state. Metcafe in his letter to John Adams had mentioned that almost 10 per cent of such Maratha demands were never met and normally the Marathas wrote off 20 per cent of amount if demands were met in advance.[74] But this tendency of the Maratha *sardars* to demand constantly and the non-payment of complete amount or payment of only a partial amount forced the Marwar state to fall in debt trap. The cyclical effect on the economy was detrimental. In order to collect the dues, the Maratha *sardars* or their representative often visited Jodhpur and every time some new promises and arrangements were made and fresh time bound installments were fixed. But the Rajput state had to immediately pay some amount in cash known as barat *kharch* and *lashkar kharch*.

The internal feuds had given encouragement to the Maratha interference in Marwar. And the conflict of Abhay Singh with Bakht Singh for the succession of the throne of Jodhpur had made a beginning on a major scale in this regard. In 1748 Abahy Singh had asked for help from Malhar Rao Holkar against Bakht Singh and had agreed to pay Rs. 11000 per day to Holkar for his favour.[75] Then in 1749 A.D. after the death of Abhay Singly a civil war started between Abhay Singh's son Ram Singh and Bakht Singh for the succession to the throne of Jodhpur.[76] Ram Singh took help of Malhar Rao Holkar once again and from now onwards the Maratha interference in Marwar became very frequent. By that time the power of the Mughal empire had declined as well and the protection and control, which was earlier provided by the Mughal empire to the Rajput states was no more possible. The Marathas, thus, did not have any fear from the Mughal empire. The Jaipur state, which had emerged very powerful during the 18th century, also could not check the Marathas because of their own fragile and weakening authority due to internal family feuds.[77] A combined Rajput force was also not easily possible since the rulership in each of the Rajput states was based on clan solidarity or *bhai-bant*. They had once tried to present a united front against the Marathas in 1734, when the first direct interference of the Marathas in the internal affairs of Bundi started, but that too had failed.[78]

Thus, now the Marathas had got a freer hand in Rajasthan and the internal political situation in states like Marwar was also very encouraging for them. After the first battle of Merta, an agreement was reached between the Rathore ruler of Marwar, Vijay Singh and Dattajee Sindhia in 1756. Vijay Singh agreed to pay Rs. fifty lakh as war indemnity. Out of it, Rs. twenty-five lakh was to be paid within a year and the rest fifty per cent was to be given in two years. Further, Marwar was supposed to pay Rs. one lakh fifty thousand every year to the Marathas.[79] Vijay Singh also promised to keep one of his army units at Ajmer at his own expenses to help and protect Ajmer, which had now been handed over to the Marathas.[80] But all these were promises huge enough that could not have possibly been met on time. It became really difficult for Marwar because the state had been going through a phase of continuous famine for the last few years. We have reference to frequent famines from 1755 to 1759 A.D.[81] Maharaja Vijay Singh tried his level best to convince the Marathas of his inability to pay such a huge money, but they were not convinced and threatened to attack Jodhpur and give the throne to Vijay Singh's rival Ram Singh. Vijay Singh was forced to pay the dues promised in 1756.[82]

After 1758 Marwar could breathe a shy of relief when Ahmad Shah Abdali in Punjab challenged the Marathas. Vijay Singh preferred to support the Afghan ruler against the Marathas and stopped paying any money to them. However, Ahmad Shah Abdali wrote to Vijay Singh in February 1761 that he should send the tax, until then paid to the Marathas, to him and promised for military help in return.[83] But we do not get any account of such payments to Ahmad Shah Abdali by Marwar because soon after, in March 1761 A.D, he left India. The whole scenario changed once again after Abdali's departure and the Marathas started demanding their dues from Marwar. Mahadji Sindhia demanded Rs. three lakh as compensation apart from the regular Rs. one lakh fifty thousand as *mamlat.* And their demands were met immediately after a threat.[84]

These huge regular payments to the Marathas left the treasury of Marwar empty. The ruler was not in a situation to pay the amount in cash and so he started making payments in terms of *hundi. In* 1765 Vijay Singh agreed to pay Rs. ten lakh as *hundi.* On many occasions the Marwar ruler asked the *Sahukars* and *Mahajans* to pay the dues to the Marathas and the state used to pay such loans to these merchants later on. The Jodhpur Records of the period give reference to such payments by the *Mahajans* and *Sahukars.* In 1766 A.D. Nandvana Bohra gave Rs. one lakh ten thousand and in 1806 A.D. Kushalchand gave Rs. fifty thousand as loan to Marwar state.[85] *Maharaja Vijay Singh-ri-Khyat* also mentions that,Vijay Singh took loan from local *Rom* on high interest rate.[86] G.S.L. Derva is quite correct when he says that *Rom* (Mercantile Community) were dependable financiers and friends in times of need of the state.[87]

The primary records of the second half of the 18th century frequently mentioned about a tax called *rahdari,* which existed earlier also, but there was a perceptible change in the mode of collection now. *Rahdari was* a kind of transit duty taken from the traders who carried various commodities through the territory of Marwar. It is significant to note that during the later half of the 18th century the right to collect *rahdari* began to be given on *ijara* in lieu of a fixed amount, whereas earlier only the state officials realized it. The *rahdari* of Merta was given to Khushal Chand of Jodhpur for Rs.28401 from Magh Vadi 12, V.S.1833 to Magh Vadi 11, V.S.1834.[88] Similarly the *rahdari* of *pargana* Nagor was given on *ijara* for Rs.50001 to Modi Har Nath for period of one year, i.e. from Magh Vadi 1, V.S.1841 to Magh Sudi 15, V.S. 1842.[89] This new trend of *rahdari* collection was probably prompted by the need of the state for immediate and more revenue.

Thus, the economy of Marwar had reached such a low that the state had to look out for financial help from the Mahajans. But this was not enough for the continuous interference of the Marathas and in turn their demand for money. We have mentioned earlier in this paper that by the beginning of the second half of the 18th century the *tankhwa jagir* outside Marwar had shrinked and Abhay Singh even started collecting lump sum amount from the *pattadars* to pay the Marathas.[90] After him Vijay Singh also continued this practice and infact imposed a lump sum amount of Rupees three hundred for every one thousand *rekh* from the *pattadars* and termed it as *mutallab*.[91] In 1765 Vijay Singh further introduced a new tax called *Rakhbab*.[92]

The continuous battle fought between 1752 and 1756 had completely crumbled down the economy of Marwar. The treasury became empty. *Khalisa* land could not be cultivated during these years since the cultivators deserted the area with fear, and as a result a huge source of the income of the state was lost. Jalore *pargana* was given to Ram Singh after the treaty of 1756, where a huge treasury of the Rathors was lying for years. Ajmer, Taragarh and Sambhar also went away to the Marathas by the same treaty. These regions were a major source of income for the state through the salt duties and other trade duties on various goods. Further, the whole of south-east Marwar, which was agriculturally more fertile as compared to other regions of Marwar, went over to Ram Singh. Now Vijay Singh was left with relatively productive areas of only Jodhpur and a major part of semi-arid zone. Thus, on the one hand, the income of the state was on decline, on the other hand the expenditure of Marwar increased by leaps and bounds due to continuous war with the Marathas, internal disturbance and huge Maratha demands.

In 1769 Mahadji Sindhia wrote to Vijay Singh that he should pay all the dues and *nazrana*. He also asked him to send the *mamlat* for the next three years in advance. For such advance payment, Marwar was to get a relief of Rs. forty thousand.[93] Vijay Singh sent Rs. three lakh eighty-eight thousand eight hundred and thirty-five as the due for 1766 to 1769. He also agreed to pay Rs. five lakh ten thousand as advance *mamlat* for the years 1769 to 1772. This amount also included Rs. twenty-five thousand as *nazrana*.[94] In the year of 1769 Vijay Singh had to pay nearly Rs. nine lakh to the Marathas. But the only relief for him was that he had received Rs. fifteen lakh from Ratan Singh of Mewar for his military support in succession dispute.[95] Howerver, the Marwar ruler had not been able to pay Rs. five lakh ten thousand till 1772 and so he received a threatening letter from Mahadji Sindhia.[96] It was only in 1774-75 that Vijay Singh was able to pay this due in two installments. He paid Rs. four lakh eighty-nine thousand four hundred eighty-three and seven ana in 1774 and then in 1775 he sent Rs. sixty thousand five hundred sixteen and nine ana.[97] Hardly had he cleared the dues for 1769-72 A.D. that another demand for 1772-77 was put forward before him. In 1781 A.D. Vijay Singh managed to pay Rs. nine lakh as *mamlat* for Marwar at the rate of one lakh eighty thousand per annum. He also paid Rs. seventy-five thousand as tax for Godwad. In 1784 A.D. Mahadji Sindhia sent his demand letter for the payment of *mamlat* of 1778-82, which stood as Rs. nine lakh. After a rebate of Rs. two lakh Vijay Singh was supposed to pay Rs. seven lakh. However, he could pay only Rs. two lakh in 1784 due to lesser revenue collection in 1782 owing to famine, leaving outstanding dues of Rs. five lakh. Before this due could be paid by Marwar another demand of Rs. seven lakh fifty thousand was put forward as *mamlat* for the years 1783-88 A.D. Due to fear of attachment the state paid two installments of Rs. one lakh and Rs. two lakh in 1785 A.D. to the Marathas. Then they paid Rs. one lakh fifty thousand in 1786 AD.

Now it had become impossible for Marwar to bear the unending burden of taxation by the Marathas. By that time Mewar and Jaipur were also fed up with the Maratha interference in the internal affairs and their continuous huge demands for *peshkash, mamlat* or *nazrana*. All the Rajput states willingly or unwillingly got united to fight the battle with the Marathas.[98] A battle was fought at Tunga in July 1787.[99] The Marathas under Sindhias lost the battle and the Rathors were able to capture Ajmer and the surrounding regions. Now Marwar stopped paying any money to the Marathas. But this could become possible only for three-four years. Mahadji Sindhia attacked Marwar in 1790 (first in June and then in September). The battle at Merta was fought and Marwar lost the battle. The defeat proved disastrous for the Rajput state since the damages claimed after this battle by the Marathas were quite heavy.[100] The following were compensation for the loss and the damage suffered claimed was :[101] (a) war indemnity, *nazrana*, *darbar kharch* and *khasa sawari* of Rs. sixty lakh. Out of these Rs. fifteen lakh was to be paid by Marwar in cash, Rs. eight lakh was to be given on 4th February 1791 and the rest by 4th April 1791. *Bharna,* which included jewellery and horses etc., was fixed at Rs. three lakh while Rs. fifteen lakh was to be given for the distribution among the Maratha army along with Rs. three lakh worth of *bharna.*

Vijay Singh was asked to pay Rs. four lakh *bharna* in equal instalments within two years. He was also supposed to pay Rs. twenty lakh in four years time as annual tax for Navan, Maroth and Parbatsar. Mahadji also mentions in his letter to Vijay Singh that if all these money is paid on time he shall give a rebate of Rs. two lakh. (b) Vijay Singh was also forced to agree on the terms of paying Rs. one lakh fifty thousand as the annual tax of Marwar and Rs. thirty thousand as annual tax of Godwad. Apart from this, Marwar also had to cede certain rights of territories like Sambhar, Kherva, Masuda and Bhiay and were given on *ijara* to the Marathas.

As the terms and conditions of the post-Merta war show it was simply impossible for Marwar to manage such a huge amount for the Marathas. A new trend of *ijara* being given to the Marathas started which meant maximum exploitation of the peasantry and in turn of the land.[102] The state under Vijay Singh had fallen in a trap where there was no escape and the only hope was that the Marathas themselves opt to get out. But this could not be done by the end of the 18th century. In fact, it is only after the Marathas started facing the challenge from the British and to a large extent because of amplified rivalry among the Maratha *sardars* that this practice of hopping collection of money decreased.

Impact on Land-Use Pattern

The regular payments to the Marathas forced the state to generate more revenue and that was only possible either through invading a new territory and bringing in war-booty or extracting more revenue from the agrarian and non-agrarian sectors. Since the state was not in such a strong position militarily, the possibility of expansion of the territory is ruled out. The option left before the state was to introduce some new taxes to be paid by the cultivators and the traders. Though the state made immediate gains by the collection of these revenue, specially land revenue, but the pressure of revenue was so much that many of them deserted the area. This in turn affected the production process. The need for greater land revenue began to be felt by the end of the 18th century and the state started making all efforts to bring back the cultivators who had earlier deserted the area. We have reference to such account primarily from neighbouring territories of

Table 51.4 : Demand and Payments to the Marathas by Marwar State

Year	*Demand*	*Payment*	*Remarks*
1756	Rss. 50 lakh+ Rs.1.5 lakh every year	Rs. 25,00,000	Promised to pay the rent.
1757	Rs.12.51akh	Rs.12,50,000	
1758	Rs.12.5 lakh	Rs. 12,50,000	
1763	Rs. 1.5 lakh + Rs. 3 lakh as war indemnity	Rs. 4,50,000	
1765	All the dues at the rate of Rs.1.5 lakh per annum	Rs. 10,00,000	In *Hundi*
1769	All the dues and *Nazrana*	Rs. 3,88,000	Agreed to Rs. 5.1 lakh for 1769-72.
1772	Rs. 5.1 lakh + Rs. 25,000 as revenue of Godwad and Rs. 5,000 *Nazrana.*	Not paid	
1774	All the dues	Rs. 4,89,483	
1775	Rest of the dues + *Mamlat* for 1772-77.	Rs. 60,516	
1778	All the dues	Rs. 30,500	
1779		Rs. 1,00,000	
1780		Rs. 69,000	
1781		Rs. 9,00,000 Rs. 75,000	
1784	Mamlat for 1778-82 Rs. 7 lakh after rebate of Rs. 2 lakh	Rs. 2,00,000	
1785	Mamlat for-1783-88 Rs. 7.5 lakh	Rs. 1,00,000 Rs. 2,00,000	Stopped paying after defeating the Marathas at Tunga.
1791	Rs. 60 lakh under different heads *Mamlat, Nazrana, Fauz Kharch, Darbar Kharch, Khasa Sawari.*	Rs. 4,00,000	Rest to be paid in instalments
1792	Rest of the dues	Not mentioned Rs. 14,00,000 (Average payment of Ks. 2 lakh a year)	Contemporary accounts mention that regular payments as promised were made.
1800	Rs. 6 lakh	Rs. 2,00,000	
	Total payment	Rs. 1,12,26,499	

Note : Average for 45 years = Rs. 2,50,000 approximately.

Marwar. One such account is of 1796 A.D. when Maharao Ummed Singh of Kota sent a *parwana* to the *patel* of Jodhpura village in *pargana* Madhukargarh and asked him to invite peasants from other villages to settle in his village. Another *parwana* was sent in 1804 A.D. to the *patel* of Bhumbhari Richwa village to bring the entire land of the village under cultivation.[103]

Apart from the non-regular extra taxes from the cultivators, traders and *thikanedars*, Maharaja Vijay Singh of Jodhpur also tried to bring in some money from outside the state, but there is a solitary instance from Mewar. In 1769 A.D. Vijay Singh got an opportunity in Mewar where Ratan Singh had revolted against Maharana Ari Singh. Few of the feudatories, who were supporters of Ratan Singh wrote to Vijay Singh for help and for this they promised to pay Rs.15 lakh to him. Since the Marathas were also supporting Ratan Singh, Vijay Singh thought it wise both strategically and financially to help Ratan Singh against Ari Singh of Mewar and earned Rs.15 lakh out of it. However, Vijay Singh had to immediately pay nearly Rs.9 lakh to Mahadji Sindhia that was lying as outstanding due on Marwar.

We have mentioned earlier that Abhay Singh had introduced a new practice of collecting lump sum amount from *pattadars* to pay huge amount in turn to the Maratha. After his death Maharaja Vijay Singh in 1755 A.D. started collecting a fixed lump sum amount of Rs. 300 for every 1000 *rekh* from the *pattadars* called *Mutallab*.[104] Another new development of the 18th century after the Maratha raids was the introduction of the *ijara* system by the state. The *ijaradars* were supposed to pay Rs. 650 on the *rekh* by Rs. 1200 in *Khalisa* lands.[105]

Since we are not in access of exact *hasil* figures of the 18th century, some hypothetical position is presumably possible by extrapolating our analysis on the basis of the scattered data for the 18th century. We are in access of *rekh* for the year 1660 A.D. We have mentioned above that the pattadars started praying Rs. 300 for every 1000 rekh after 1755 A.D. So keeping the least possible *rekh* amount of Rs. 25,86,941 in consideration the extra burden on *pattadars* comes to roughly Rs. 8,06,231. It is this minimum amount that the *pattadars* have to extract more from their territories. In other words, Rs. 8,06,231 is an extra burden on peasantry and if we add the figure to *hasil* it comes to Rs.33,30,790, which is, much more than *jama* figure. As discussed earlier also, since *jama* was the actual maximum capacity of the peasantry to pay, anything more than that might have forced them to make intensive use of land, and if the situation did not improve possibly it had led to land degradation. In the absence of new agricultural technology the fertility of the soil might have been affected. The following diagram can be prepared on the basis of the above discussion.

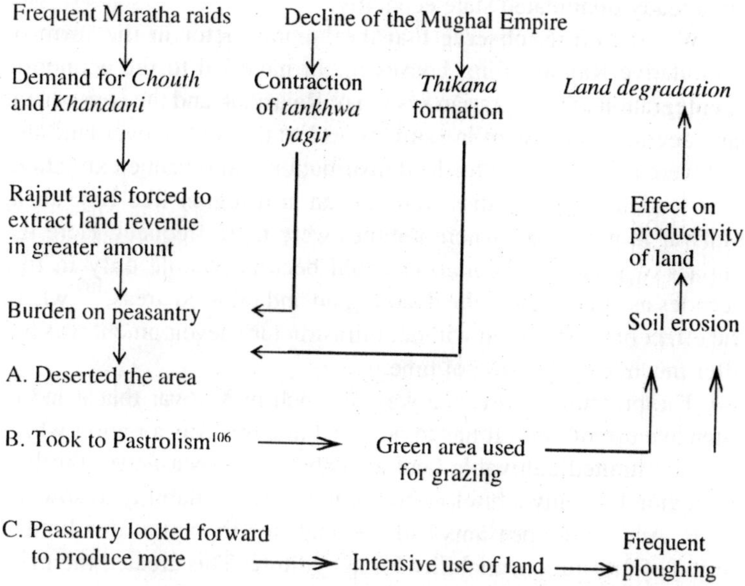

An important thing to understand here is that the state had to anyhow collect an average amount of Rs.2 to 3 lakh per annum for the extra payment it had to make to the Marathas. The ruling authority of Marwar began to work out more ways to collect extra revenue from the peasantry. The following feelings of one of the Rathor *sardars* is an ample proof of that: '......सिंहओ खूबचन्द अरजइ किवी के मूलक में बाब घालिया खरची पुगसी'।[107] It means that special taxes will only fulfill the need. The economic pressure was so much on the state that the center of power at Jodhpur began to feel weak and the *pattadars* started defying orders of the state and began collecting money from the peasantry forcefully. The following version in one of the contemporary accounts makes the picture more clear : '... सु विशनसिंह सु मिल उना ने मिल पकड रुपिया बावन हजार माथे किया, जना दिया काना सु। फेर रुपिया लिया सो एक लाख रुपिया पेदस किया। ने सवाइ रामजी ने गोरधन जी ने कवायो के थे बैठा ही मारो पडकर खोसे है।...[108] Such forceful extractions became very common in Marwar during the second half of the 18th century. All these aspects are possibly pointer to the fact that despite increased production and increase *hasil* (actual realization) the pressure on peasantry increased with the passage of time due to greater Maratha demands, worsening the already dilapidated state economy.

We, therefore, observe that the human factor in the form of exploitative Rajput ruling houses might have led to the economic disintegration of the agrarian system of Rajasthan and the consequent land degradation. The ruling authorities lost the control over land-use and were anxious more for land distribution and revenue extraction. This tendency proved disastrous for an arid region like Rajasthan, Marwar in particular, where famines were more frequent, Here the impact of greater exploitation could become visible only in few decades as compared to the doab region and rain-fed areas,[109] where the effect of exploitation without infrastructure development was felt after much longer period of time.

The pressure of revenue was so much in Marwar that it led to intensive use of land. It has to be noted that this was a region where one had limited cultivable land available to the peasantry. Further, the region had only a brief period of moisture availability to sow the seeds. And so the peasants had the tendency to exploit this limited period of moisture availability for agriculture. This created the main erosive hazard to the fertility of the soil because of human activity. Since the peasants were under constant pressure, they had no option but to either till the soil or to desert the area. The moisture less soil was disturbed by these 'forced ploughing' and because of scarcity of water, surface soil used to get dried out and led to greater wind erosion of the soil. Soil is one of the main processes of land degradation and consists of physical detachment of soil particles by wind. This was a very common feature of Rajasthan and it led to the deterioration of the overall productive potential in the soil. In turn, soil degradation resulted in decrease of the productive land, ultimately forcing the farmers to encroach upon unsuitable and vulnerable lands—wastelands, slopes and other marginal areas, and the 'village commons'. It is but natural that per capita productive land and soil quality rapidly diminished and the cycle of soil degradation and the pauperization of peasantry continued even faster.[110]

During the period of our study the cultivators made more and more constant use of land for cropping and thus disturbed the earlier practice of leaving it fallow for replenishing its fertility. This resulted in the use of land beyond its ecological capabilities and as a consequence its productivity must have been affected.[111] Further, in Marwar there was a tendency to shifting cultivation and during the 18th century there was shrinkage in this practice as well.[112] The settlement Report of 1930s also say that after three years of cropping, the land was left fallow for the next three years indicating that this was a common practice in the arid zone of Marwar.[113] We must note that during the 18th century when pressure of revenue increased due the frequent Maratha demands the concept of leaving the land fallow was gradually being forgotten and so the impact on the fertility of the soil was negative. Contrary to the general practice of the cultivators to leave the land fallow to regain its fertility, there was another kind of shifting cultivation under practice called *walar* system. In this system of shifting cultivation there was a direct human assault on forested area by the Bhil and Girasia tribes. They used to cut and burn the jungle and sowed the seeds in the ashes. This method of cultivation was also destructive for the fertility of the soil. But here since we are mainly concerned with the fallout of the Maratha incursion we shall not go into the details of such human factor in land degradation.[114]

We are not in access of detailed statistical data for the 18th century Marwar and so we are not able to work out exact area of land use for cultivation. But we do have figures for the second half of the 17th century and the end of the 19th century for different *parganas* of Marwar. A comparison of the two periods would reflect the trend of the continuation or changes if any of the land use pattern. Let us evaluate the figures of the two periods for *pargana* Merta of Marwar.

Table 51.5 : Land-use in *Pargana* Merta[115]

Year	*Cultivable Land*[116]	*Uncultivable Wasteland*	*Fallow*	*Total Measure (in Bigha)*	*Primary Source*
1572-73	N.A	N.A.	N.A.	21,14,773	*Ain-I-Akbari*
1660[117]	18,11,347	1,62,833	N.A.	20,33,726	*Marwar-ra pargana-ri-Vagat*
1898-99	18,78,420	3,22,620	6,98,500	22,01,040	Agricultural Statistics of India Series

Table 51.6 : Uncultivable Wasteland in Pargana Merta

Different wastelands	*Area in bigha-I- Sikandari*
Gaon khere (habitable)	12721
Nalla khohla (water drainage)	15687
Nadi (river)	13341
Rah (road)	2032
Thali (desert)	11921
Haudach (pasture ground for elephants)	45785
Sorkalar (salty or saline land)	24138
Sigdanto (pasture land for cattle)	3725
Pahari (hilly area)	36399
Total	165749

If we compare the two tables we find that Munhata Nainsi's own figure of uncultivated wasteland (Table 51.5) is contradictory to the total of the breakup of the different wastelands given by him. We may presume that he has missed some category of wasteland while giving details. The difference comes to 49681 *bigha* (*Sikandari*). Probably this could have been shrub and grass area, which is generally found in abundance in arid and semi-arid zone like Marwar. Further, there is not much of a difference in the total measured area from 1572-73 down to 1898-99. There is also not any major increase in the cultivable land area. This proves our earlier assumption that maximum possible area was brought under cultivation during the Mughal period itself and therefore, in the absence of further scope for expansion, the peasantry had to make intensive use of the available cultivable land beyond its productive capacity. It has also been highlighted by B.L. Bhadani with evidence of *pargana* Sojhat that the area was fully cultivated.[118] In fact, in the whole of Marwar the situation was almost the same with possibility of slight variation. The state of cultivation here during the 17th century and even in the 18th century was not very different in terms of area tilled from what existed in the first half of the 20th century.[119]

It is important to mention here that after the decline of the Mughal empire many other regions (*Suba*) also went through a phase of economic decline and naturally the pressure was felt most by the peasantry. But in these *Suba* as compared to Rajasthan, specially Western Rajasthan, fertility of the soil was not affected. When the burden became heavier on the peasantry, they either revolted or chose to desert the area and moved to some other region or even brought much new barren land under cultivation.[120] Seemingly, the situation of the Mughal territory even during the heydays was not very conducive for the peasantry. The exploitation of the surplus produced was always there. The state was never lenient in collecting revenue from the peasantry. However, the situation in Rajasthan as a matter of fact was totally different. There was a limit to surplus of exploitation here. Before the 18th century the rulers were in a comfortable position financially without putting any extra burden on the peasantry and by bringing the maximum possible area under cultivation. The large income of Marwar rulers enabled them to accumulate huge treasure. Khwaja Agar, formally the *diwan of* Jaswant Singh, who finally joined the services of Aurangzeb, informed the emperor that Jaswant Singh had a treasure hoard of Rs.57 crore and Rs.1000 was deposited everyday in the fort's treasury.[121] *Ajit Vilas* also records that Jaswant Singh had left a huge treasury when he died.[122] But Rajasthan was a region which was not abundant in quality and cultivable barren land as compared to other territories of the Mughal empire. So when the pressure for revenue intensified, because the recurrent Maratha demands it was but natural that efforts were taken to extract more and more from the available cultivable land. It is pertinent to note here that land revenue had been the major source of income of the state from time immemorial. In contrast, the other areas of the Mughal empire like Awadh, Bengal, Malwa etc. were able to accommodate and adjust itself to pressure by expanding the cultivable area. The new barren land were brought under use and they acted as cushion or safety-valve and lessened the effect of over-exploitation of the peasantry through land revenue. Thus, Marwar was an area with a difference because of its geographical limitations and limited quantity of cultivable land. And any extra burden on the peasantry meant forcing them to cultivate more and produce more and it might have affected the fertility of the soil as well. But this should only be taken as presumption since it is impossible to get any data on the fertility of the soil over a period of time during the 18th century.

To conclude the discussion, the Maratha *sardars* did not put forward their demands directly to the cultivators or the traders. The primary records give reference to rare instances of forceful collection by the Marathas from the cultivators and the traders. We do get account of forceful collection of grains from the villages that fell on the route taken by the Maratha *sardars*. But generally speaking they in common did not loot the people. The records show that there are only a few instances of the traders being looted by them. In fact, the Maratha ravages were mainly confined to the realization of *chauth, mamlat* and *nazrana* from the rulers. The trade *carvans* were rarely the target of the Maratha invaders in Rajasthan.[123] But in the ultimate analysis the Maratha demands did affect the economy of Marwar adversely as well as the ecology of the region.

NOTES

1. William Franklin, *Military Memoirs of George Thomas,* London, 1805, p.347.
2. Col. Tod, *Annals* and *Antiquities of Rajasthan,* vol.I, Cakcutta, 1916, p.1.
3. see Irfan Habib's *An Atlas of the Mughal Empire,* Oxford University Press,. Delhi, 1982, p.viii.
4. The Sardar Museum, MS, Jodhpur.
5. W.W.Hunter, *The Imperial Gazetteers of Rajputana*, Rajputana Province, Oxford, 1921, p.1.
6. G.N. Sharma, *Social life in Medieval Rajasthan (1500-1800 A.D.)*, Lakshmi Narain Aggarwal Educational Publishers. Agra, 1968, p.1.
7. 'Rajasthan as a Region: Myth or Reality' in Karine Schomer, Joan L. Erdman, Deryck O. Lodrick and Lloyd I. Rudolph (eds.), *The Idea of Rajasthan: Explorations in Regional Identity*, Vol. I, Manohar, New Delhi, 1994 p.2.
8. SS. Dhabriya, *Eco-Crisis in the Aravalli Hill Region*, Environmentalist, Jaipur. 1988, p 1.
9. G. N. Sharma, *Social Life in Medieval Rajasthan (1500-1800)*, Agra, 1968, p. 2.
10. *India : Case study on Desertification,* Luni Development Block, United Nations Conference on Desertification, August-September 1977, Central Arid Zone Research Institute, Jodhpur, 1977.
11. L.A. Ramdas, 'Monsoon and Rainfall Pattern in Indian Subcontinent', in B.C. Law (ed.), *Mountains and Rivers of India,* National Committee for Geography, Calcutta, 1968.
12. Carol Henderson, 'Famines and Droughts in Western Rajastan : Desert Cultivators and Periodic Resource Stress,' in Karine Schomer and others (eds.), *op. cit.,* Vol. II, p.4.
13. K.S. Singh, *People of India:Rajasthan,* Vol. xxxviii, Part I, Anthropological Survey of India. Popular Prakashan, Mumbai, 1998, p.1.
14. S.S. Dhabriya, *Eco-Crisis-the Aravalli Hill Region,* p.7.
15. Karni Singh, *The Relation of the House of Bikaner with the Central Powers (1465-1949),* Munshiram Manohar Lal, New Delhi, 1974, p.6.
16. RK. Sharrna, *Maratha Relations with Major States or Ratputana*, N. Delhi, 1973, p. 18.
17. R.C. Sharma, *Settlement Geography of the Indian Desert*, New Delhi, 1972, pp. 7-10.

18. B.L. Bhadani, *Peasants, Artisans and Entrepreneurs: Economy of Marwar in the Seventeenth Century*, Rawat Publication, Jaipur, 1999, pp. 20-27.
19. *The Ruling Princes, Chiefs and leading Personages in Rajputana and Ajmer*, Govt. of India c.f. R.K. Saxena, *Maratha Relations with Major State of Rajasthan*, p.18
20. G.D. Sharma, *Rajput Polity, A Study of Politics and Administration of the State of Marwar*, 1638-1749, Manohar, New Delhi, 1977, p.3.
21. *Ibid.*, p. 5.
22. Narain Singh Bhati (ed.), *Marwar-ra-Pargana-ri-Vigat* by Munhat Nainsi, Vol.II, Rajasthan Archyavidya Pratishthan, Jodhpur, 1969, pp 40-42.
23. *Ibid.*, Vol. I, 1968, p.38.
24. Munhta Nainsi, *Marwar-ra-Pargana ri-Vigat*. Narain Singh, Bhah (ed.), Vol.I, 1968, pp.42-44.
25. Akbar captured Merta in 1562. Abul Fazl, *Akbarnama,* Maulawi Abdur Rahim (ed.),Vol.II, Calcutta. 1879, pp. 161-62.
26. S. Nurul Hasan, *Thoughts on Agrarian Relations in Mughal India,* New Delhi, 1971. p. 37.
27. B. L. Bhadani, *Peasants, Artisans and Entrepreneurs: Economy of Marwar in the Seventeenth Century,* Jaipur, 1999, p. 2.
28. Irfan Habib, *Agrarian System of Mughal India,* Bombay. 1963, p.164.
29. *Vir Vinod,* Vol.II, Udaipur, 1886, p. 815. Also see V.S. Bhargawa, *Marwar and the Mughal Emperors,* Delhi, 1996, p.77.
30. *Ibid,* p. 815.
31. *Marwar-ra Pargana ri-Vigat.* Vol.I, p.77.
32. *Ibid.*, pp.95-106.
33. G.D. Sharma, *Rajput Polity, op. cit.* pp. 49-53.
34. Shivduttdan Barhat (ed.), *Ajit Vilas*, Jodhpur, 1984, p. 28.
35. *Ibid.* p.30.
36. G.D. Sharma, *Rajput Polity,* p. 169.
37. *Jodhpur Rajya-Ki-Khyat* (MS), no. 199, p. 56, Anup Sanskrit Library, Bikaner.
38. *Ajit Vilas*, p.49.
39. *Ibid.*, p. 52.
40. *Ibid.*, p. 65.
41. c.f. G.D. Sharma, *op.cit.*, p. 179.
42. *Ajit Vilas,* p. 71.
43. Irfan Habib in *Agrarian System of Mughal India,* Oxford University Press, New Delhi, 1999, gives emphasis on the crisis of the agrarian system, which led to the decline of the empire. The author believes that the structure of the empire was erected on the sound agrarian system and revenue collection but when it was disturbed, the empire gradually crumbled down. M. Athar Ali in his *The Mughal Nobility Under Aurangzeb*, New Delhi, 1997 talks about the *Jagirdari* crisis while Satish Chandra in his two books *Parties and Politics at the Mughal Court (1707-1740)*, New Delhi,1959 and *Medieval India: Society, the Jagirdari Crisis and the Village,* Madras, 1982 finds the problem in the medieval Indian economy and the *Jagirdari* system under later Mughals. Muzaffar Alam and Sanjay Subrahmanyam disagree from the above views. However, they feel that rather than insisting too

greatly on the existence of a single logic that informs Mughal decline, it might be worthwhile to focus on the vast variety regional experiences. See *The Mughal State 1526-1750*, Oxford University Press, Delhi,1998, p.59.

44. Satish Chandra, *Parties and Politics at the Mughal Court*, p.186.
45. Cited in *ibid.*, p.186.
46. Dilbagh Singh. *The State Landlords and Peasants: Rajasthan in the 18th Century*, New Delhi, 1993, p. 198.
47. Satya Prakash Gupta, 'Reconstructing the Political and Economic Profile of Rajasthan', Address of the Sectional President, *Proceedings of the Indian History Congress*. 55th Session. Delhi, 1995, p.167.
48. See Vipul Singh, 'Professional Mobility and Change: The World of Artisans in the Late 18th Century Patna', *Studies in History and Culture,* Vol. 4, No. 2, 1996, pp. 31-37
49. Sukumar Bhattacharya, *The Rajput States and the East India Company,* New Delhi, 1972.
50. Dilbagh Singh, *op cit.,* p. 199.
51. Marcia J. Frost,'Coping with Scarcity: Wild Woods and Common Land. Kheda District (Gujrat, India), *1824/5', Indian Economic and Social History Review (IESHR),* Vol. 37, No. 3. July-September 2000, p. 311.
52. The Rathor *sardars* who had hereditary claim in their territories of rule were known as *thikanas* and the owner *of* these *thikanas* were known as *thikanedars.*
53. R.H. Grove, *Green Imperialism: Colonial Expansion. Tropical Island Edens and the Origins of Environmentalism (1600-1860)*, Cambridge, 1995.
54. G.D. Sharma, *Rajput Polity*, p. 251.
55. *Vir Vinod*, pp. 1218-21.
56. *Marwar-ra-Paragana-ri-Vigat,* Vol. II, p. 51.
57. Satish Chandra, *Mughal Religious Policies, the Rajputs and the Deccan,* New Delhi, 1993, p. 34.
58. S. Inayat Ali Zaidi; The Mughals and the Rajputs 1605 - 1659 AD., Ph.D. thesis, Aligarh Mulsim University 1982, pp. 200-237, has discussed at length the whole concept of *watan jagirdar*. He has shown *watan jagir* gradually displaced the past concept of the Rathor of *bhai-bant,* which was not transferable unlike the normal *jagir* given by the Mughals.
59. The use of patta as revenue right was confined to Western Rajasthan. In Eastern Rajasthan the Mughal term *jagir* was used. See S.P. Gupta, *The Agrarian System of Eastern Rajasthan (C.* 1650-1750), Delhi, 1986, pp. 211-212.
60. For details of the *mansabdari* system see Irfan Habib's article 'The Mansab System (1159-1637)' in *Proceedings of Indian History Congress*, 1967. Also see M. Athar All, The Mughal Nobility Under Aurangzeb, OUP, New Delhi, 1997; and Satish Chandra, *Parties and Politics at the Mughal Court,* 1707-1740. New Delhi, 1972 pp. XXVII-XXVIII.
61. The *rekh* began to be mentioned in *pattas* since the time of Mota Raja Udai Singh (1538-95). For more information see Satish Chandra, Raghubir Sinh and G.D. Sharma (eds), *Marwar Under Jaswant Singh: Jodhpur Hakumat-ri Bahi (1658-781),* Meerut, 1976, pp. 125-237.
62. B.L. Bhadani, *Peasant, Artisans and Entrepreneurs*, p. 262.

63. See G. D. Sharma, *Rajput Polity,* p. 91.
64. The primary accounts mention *mal* as some kind of tax other than land revenue B.L Bhadani believes that it must have been some kind of irrigation tax. See his *Peasants, Artisans and Entrepreneurs,* p. 218. G.D. Sharma, *Rajput Polity*, pp. 86-88 and 101 says that *mal* was nothing but land revenue. But for land revenue a term called *bhog* has specifically been used in *Marmar-raPargana-ri-Vigat,* Vol. II, pp. 95-97, and *mal* as land revenue was used in Mughal administration as *mal-oljihat.* See Shireen Moosvi, *The Economy of the Mughal Empire, C. 1595, A Statistical Study*. Delhi, 1987, p 129.
65. *Marwar-ra-Pargana-ri -Vigat*, Vol.ll, p. 89. Also see for a better analysis G.D Sharma, *op. cit.*, pp. 100-101.
66. *Sanad-Parwana-Bahi.* No.13. V.S.1830 (1773A.D.), Jodhpur Records, Rajasthan State, Archives, Bikaner.
67. *Annals and Antiquities of Rajasthan.* Vol. II, p.160.
68. *Sanad-Parwana-Bahi,* No.6, *Asadh Vadi* 2, V.S.1874 (1767A.D.), f.183. Jodhpur Records, R.S.A. Bnkaner.
69. *Sanand-Parwana-Bahi* No. 14. V.S.1831 (1774A.D.), f.92, Jodhpur Records, R.S. A. Bikaner.
70. Dilbagh Singh, *The State, Landlords and Peasants: Rajasthan in the 18th Century,* 1990, pp.198-99.
71. *Ibid.*, p. 204.
72. *The Selections from the Peshwa's Daftar*, Vol. 21, No. 82. Vijay Singh agreed to pay 1.5 lakh every year as *mamlat* is also proved by a "Letter from Madhavrao Sindhia to Mahraja Vijay Singh", on Jyestha Sudi 5, V.S. 1826 (1769 A.D.). See *Jodhpur Rajya Patra Vyavhar, V.S. 1769-2003,* Vol. I, Rajasthan State Archives, Bikaner, n.d., pp. 7-8 and 82.
73. "Letter from Mahadji Sindhia to Maharaja Vijay Singh", Paush Sudi 1, V.S. 1847 (5th January 1791), Portfolio File 6. Letter No. 67, Dastari Record, Jodhpur.
74. "Letter from Metcafe to John Adams", 15th January 1818, *Foreign Secret Proceedings, 6th February 1818*, No. 182, National Archives of India, New Delhi.
75. *Marwar-ri-khyat,* Vol. II, Anup Sanskrit Library, Bikaner, p.160.
76. *Marwar ri Khyat,* Vol. II, p. 172, and *Dayaldas ri-khyat,* Vol.ll, Anup Sanskrit Library, Bikaner, pp. 72-73.
77. See Dilbagh Singh, *op. cit.*, p. 8.
78. The queen of Rao Raja Budh Singh had sent her envoy to Poona for hiring the Marathas to oust Dalel Singh from Bundi, who was installed as the ruler by Sawai Jai Singh. See Surya Mal Mishra, *Vamsha Bhaskhar, Jodhpur,* 1899, p. 1922. Raghubir Sinh says that she got the help of Malhar Rao Holkar (then Incharge of districts north of the Narbada) and Ranoy Sindhia (the founder of Sindhi dynasty) by paying them Rs. 6 lakhs, *op. cit.*, p. 163). The two Maratha leaders brought their hordes to Bundi and captured it on 22nd April 1734 and Budh Singh was proclaimed ruler. The Kachwahi queen of Budh Singh tied Rakhi on Malhar Rao Holkar's wrist. It is significant to note here that on the one hand the queen accepted Holkar as her brother, Budh Singh on the other hand, did not even come to meet

Holkar. This shows that *the Rajputs were not ready to accept the Marathas as equals*. Soon after the Marathas left Bundi, a Jaipur force arrived there and restored Dalel Singh to the throne. See M.L. Sharma, *Kota Rajya Ka Itihas*, Vol. I Kota, p. 366. This first penetration of the Marathas into Bundi alarmed the Rajput Rajas of Rajasthan, specially Sawai Jai Singh. He planned to summon a conference of the leading Rajput Rajas. Though it was called by Sawai Jai Singh but Maharana Jagat Singh of Mewar was given the honour to preside over its deliberations. The conference was attended by Jagat Singh of Mewar, Sawai Jai Singh of Jaipur, Abhay Singh of Jodhpur, Durjan Sal of Kota, Jorawar Singh of Bikaner, Dalel Singh of Bundi, Gopal Singh of Karauli, Bakht Singh of Nagor, and Raj Singh of Kishangarh. The conference was held at Hurda, on the outskirts of Mewar, in July 1734. The Rajputs took an oath to unite their forces to resist all Maratha incursions. See Shyamal Das, *Vir Vinod*. Udaiur Raj Yantralaya, pp. 1220-21. But the Rajputs were not able to present a united front because of their internal jealousies. The four Rajputs of Jaipur, Jodhpur, Kota and Kishangarh, however, came forward and asked the Mughal court to launch a campaign against the Marathas and promised their support for it. See G.S. Sardesai, *Selections from the Peshwa Deftar*, Govt. Press, Bombay, Vol. XIV, Letter 21 & 23, 1927-32. The Mughal emperor deputed Khan-I-Dauran to lead the Mughal forces towards Malwa and he was assisted by the Rajput Rajas as promised earlier. However, the combined forces could not check the advance of the Marathas. Sawai Jai Singh, the Subedar of Malwa, was forced to grant them Rs. 22 lakhs as the *Chauth* of Malwa in 1735. This made has enemies at the Mughal court to remove him from the *subedari* of Malwa.

79. *Selections From Peshwa Dafter*, Vol. 21, No. 82.
80. The Marathas got Ajmer, Garh Bitli (Taragarh) and the surrounding territories in 1756. See 'Letter from Jankoji to Vijay Singh', Asadh Badi 16, 1813 V.S. (26th June 1756), *Portfolio File No. 108-112*, Dastari Record. Jodhpur.
81. Hukam Singh Bhati (ed.), *Marwar-ri-Khyat*, Jodhpur, 2000, p. 41.
82. Letter from Anand Rao Bavle, Kartik Sudi 12, 1815 V.S. (12th November 1758), *Arzi Bahi*, No. 4, p. 90; *Selections from Peshwa Daftar*, Vol. 2, No. 94, 95, 96.
83. G.R. Parihar, *Maratha-Marwar Sambandh* (*1724-1543 A.D.*), Jaipur, 1977, p. 72.
84. "Letter from Vijay Singh to Mahadji, Kartik Badi 4, 1820 V.S. (26th Oct. 1763), *Arzi Bahi*, No. 4, Jodhpur, p.24: *Marwar-ri-khyar*, Vol. III, pp. 34-37.
85. *Arzi Bahi*, No. 4, p. 286. Also see *Sanad-Parwana-Bahi*, 1824 V.S (1767 A.D), Jodhpur Records, R.S.A. Bikaner.
86. Brajesh Kumar Singh (ed.), *Maharaja Vijay Singhji-ri-khyat*, Jodhpur, p. 72.
87. 'Land Control and Agrarian Mercantile Classes in Western Rajasthan C. 1650-1700', in the *Proceedings of Indian History Congress*, 58th session (Banglore), Aligarh, 1998, p. 371. Also see Dilbagh Singh 'The Role of Mahajans in the Rural Economy of Eastern Rajasthan during the Eighteenth Century, *Social Scientist*, May 1974; and S.P. Gupta, *The Agrarian System of Eastern Rajasthan*, Delhi, 1986, pp. 134-44 for the role of *Mahajans* in the economy of Rajasthan.

88. *Sanad-Parwana-Bhahi, N*o.17, V.S.1833 (1776 A.D.), f.306.
89. *Sanad-Parwana-Bahi,* No.30, V 5.1843 (1786A D.), f.334.
90. *Vir Vinod*, pp. 1218-21.
91. Jodhpur Records. *Basta* No. 92/7.
92. *Maharaja Vijay Singhji-ri-khyat,* p. 72; *Marwar-ri-khyat,* Vol. III, p. 41.
93. "Letter from Majadji to Vijay Singh". Jyestha Sudi 5. 1826 V.S (8th June 1769). *Portfolio File No. 6,* Letter No. 8, Dastari Record, Jodhpur.
94. "Letter from Mahadji to Vijay Singh", Jyestha Sudi 5, Letter No. 6, *ibid.* The letter mentions the receipt of the amount by Mahadji Sindhia.
95. *Jodhpur Rajya Patra Vyavhar,* V.S. 1769-2003, Vol. I, Rajasthan State Archives, Bikaner n.d., p. 27.
96. "Letters from Mahadji to Vijay Singh". Chaitra Badi 5, 1829 V.S. (23rd March 1772), Letter Nos. 18-19. *op.cit.*
97. "Letter from Mahadji to Vijay Singh", *Portfolio file No. 6*, Leller Nos 29-30.
98. 'Letter from Mahadji to Vijay Singh', Shravan Badi, 1839 V S (26th July 1782), *Portfolio file No. 6.* Letter No. 44, Dastari Record, Jodhpur.
99. See "Letters from Gyanmal Muhanot to Zalim Singh", Bhadrapad Badi 12, 1844 V.S. (8th September 1787), *Arzi Bahi No. 4,* p. 221, Jodhpur.
100. For details of the battle see *Maharaja Vijay Singhji-ri-Khyat*, pp. 105-112.
101. The details of the terms are in 'Letter from Mahadji to Vijay Singh', Paush Sudi 1, 1847 V.S. (5th January 1791). *Portfolio File 6.* Letter 57, 58 and 59. Dastari Record, Jodhpur.
102. However, it was Abhay Singh under whom *ijara* system started. See Prem Singh Angris, *Maharaja Abhay Singh,* Jodhpur, 1987, p. 149.
103. Cited in B.L. Gupta, *Trade and Commerce in Rajasthan during the 18th Century,* Jaipur,1987, p.43.
104. *Jodhpur Records, Basta no.* 92/7, Rajasthan State Archives, Bikaner. In Basta no. 97/2 of Jodhpur Records there are references to the amount realized from various *pattadars* by Abhay Singh.
105. Cited in G.D. Sharma, *Rajput Polity*, p. 260.
106. S.P. Malhotra and H.S. Mann in "Desertification and the Organization of Society" have linked up pastrolism with land degradation in Brian Spooner and H.S.Mann (eds.). *Desertification and Development,* London, 1982, pp.306-308. Edward S. Haynes has also tried to visualize the impact of pastoralism in "Land Use, Natural Resources and the Rajput State, 1780-1980", Rakesh Hooja and Rajendra Joshi (eds.), *Desert, Drought and Development*, Rawat Publications, Jaipur, 1999, p.78.
107. *Maharaja Vijaysingh-ri-Khyat,* p.114.
108. *Ibid.*, p. 93.
109. Chetan Singh's *Natural Premises: Ecology and Peasant Life in the Western Himalaya, 1800-1950,* Delhi, Oxford University Press, 1998, is one such book which shows as to how environment was affected by human activity over a sizeable period of time in the Western Himalayan Region during the 19th century.
110. Tapeshwar Singh, 'Challenges of Land Degradation and Appropriate Conservation Measures', in R.K. Rai *et al.* (eds.) *Environment, Resources and Development,* Shillong, 2001, p. 53.

111. See S.P. Malhotra's analysis in 'Man and Desert' in Rakesh Hooja and Rajendra Joshi (eds.), *Desert, Drought and Development,* Jaipur, 1999, p.26.
112. B.L. Bhadani, Peasants, *Artisans and Entrepreneurs: Economy of Marwar in the Seventeenth Century,* Jaipur, 1999, p. 39.
113. *Settlement Report-Rent-Rate Report of Pargana Marta,* Parbatsar and Sambhar, Hawala File No.40/4, (typescript) Part I, Mehkama Khas, Govt. of Jodhpur, Rajasthan State Archives, Bikaner.
114. For Details see *Rajasthan District Gazetteer,* Vol. 2, *Sirohi,* Compiled by B.N. Dhoundiyal, Jaipur, 1967, pp. 33 and 103. Also see *Rajasthan District Gazetteer,* Vol. 14, *Dungarpur.* Compiled by K.K. Sehgal, Jaipur, 1974, pp. 8-9. The ecological destructions of Bhil Cultivation has also been reported by A.E. Lourie on Jodhpur Forests, dated 14th February 1885, *Marwar State Records,* Basta No. 1, File 3/2, Rajasthan State Archives, Bikaner, p. 22; and also by Rao Sahib Padmanabha Pillai, Chief Conservator of Forests, Marwar State, 'Note on the Kushalgarh Forests, dated 25th January 1942, *Rajputana Agency Records,* File P/28-27, 1942-44, p. 2.
115. Pargana Merta was under the imperial Mughal administration under *sarkar* Nagaur and did not belong to the state of Marwar proper during the 16th century. But during the 18th century it became part of Marwar with the decline of the Mughal Empire.
116. The figure for the year 1898-99 of the cultivable land includes cultivable wasteland and fallow land. I have deliberately calculated this figure because for the earlier period cultivable land area (figure) included fallow land. The definition of fallow land varied widely among different states of Rajasthan. See *Rajputana Agency Records*, File No. 467-6/1946, National Archives of India, New Delhi. It gives information on the current fallow in Rajputana states. The record mentions a wide range of trend of leaving the land uncultivated from Bundi's one year to Marwar three to five years. In fact, in Marwar 'long fallow' included all areas that had been under cultivation 'within memory'. Possibly it is because of this reason that the figure of cultivable land during the 17th century has been such large. It is larger than the area under cultivation in 1898-99.
117. The figure mentioned in *Vigat* is in terms of *bigha-i-sikandari,* but I have converted it in *bigha-i-Ilahi.. Bigha-i-Sikandari was* based on *Gaz-i-Sikandari* and it was replaced by Akbar in 1589 A.D. by *Gaz-i-Ilahi. Gaz-i-Sikandari* was 39/41 of *gaz-i-Ilahi.* See Shireen Moosvi, *The Economy of the Mughal Empire,* C.1596: *A Statistical Study*, OUP, Delhi, 1987,p.74. Also *see* Irfan Habib, *Agrarian System of Mughal India (1556-1707).* Bombay, 1963, pp. 355-68.
118. B.L. Bhadani, 'Population of Marwar in the second half of the 17th century', *IESHR,* Vol. XVII, Oct.-Dec.1979, pp. 415-27.
119. B.L. Bhadani, *Peasants, Artisans and Entrepreneurs*, p.41.
120. We do get reference in *Sanad Parwana Bahi* to the instructions given by the state to the *thikanedars* to bring barren land under cultivation, but they are very few since maximum possible cultivable land was in use.
121. cf. *ibid.*, p. 381.
122. *Ajit Vilas*, p. 29.
123. B.L. Gupta, *Trade and Commerce in Rajasthan During the 18th Century,* Jaipur, 1987, p. 243.

Our Contributors

A. Kumar
Ecophysiology Lab. Department of Botany, C.C.S. University, Meerut

A.K. Maji
National Bureau of Soil Survey and Landuse Planning, Amravati Road, Nagpur, India

A.R. Chandrakar
Professor of Geography, Ravishankar Shukla University, Raipur, Chhatisgarh, India.

Abu Muhammad Shajaat Ali
Associate Professor of Geography, The University of Texas at Tyler Tyler, Texas 75799, U.S.A.

Ali Mohammad
Professor of Geography, Aligarh Muslim University, Aligarh, India.

Anita Malik
Lecturer, Department of Geography, Meerut College, Meerut, India

Anupreet Singh Tiwana
Department of Geography, S.C.D. Government College, Ludhiana, Punjab, India

Aslam Mahmood
Professor and Chairman, CSRD, SSS, Jawaharlal Nehru University, New Delhi, India.

B.I. Teli
Professor of Geography, HNB Garhwal University Campus, Pauri Garhwal, Uttaranchal, India.

Chitra Sejwar
Lecturer, Department of Geography, J.S. (P.G.) College, Sikandarabad, Gautam Budh Nagar, U.P., India.

D.K. Chadha
Ex-Chairman, Central Ground Water Board, Ministry of Water Resources, Government of India, New Delhi, India.

D.K. Mishra
Department of Geography, Delhi School of Economics, University of Delhi, Delhi

D.N. Bajpai
Head, Department of Geography, Tilak College, Auraiya, Uttar Pradesh, India

D.R. Sikka
Retd. Director, Indian Institute of Tropical Meteorology, 40, Mausam Vihar, Delhi, India.

Deep Shikha
Reader, Department of Geography, R.G. College, Meerut, U.P., India.

Dharam Singh
P.G. Department of Geography, S.C.D. Government College, Ludhiana, Punjab, India.

Dirk Schaefer
Department of Geography, Research Unit for Recent Climate Change, Mainz University, Germany.

Don Funnell
Department of Geography, University of Sussex, Brighton, U.K.

G. Nageswara Rao
Central Research Institute for Dryland Agriculture, Hyderabad, India

G.G.S.N. Rao
Central Research Institute for Dryland Agriculture, Hyderabad, India

G.P. Obi Reddy
National Bureau of Soil Survey and Landuse Planning, Amravati Road, Nagpur, India

G.S. Chauhan

Department of Geography, Dr. B.R. Ambedkar College, Delhi, India.

Govind Prakash

Professor of Botany, C.C.S. University, Meerut, India.

H.M. Desarda

Ex-Member State Planning Board, Maharashtra, CIDCO, N-2A, Vithal Nagar, Aurangabad, Maharashtra, India.

Hema Yadav

Lecturer, Department of Geography, Government College, Ajmer, Rajasthan, India

Isotoli Kibami

Department of Geography, Jamia Millia Islamia, New Delhi.

Jagdish Singh

Director, Institute for Rural Eco-development, Gorakhpur, U.P., India.

Jitendra Kumar

Department of Geography, Ravishankar Shukla University, Raipur, Chhatisgarh, India

K.S. Gajbhiye

National Bureau of Soil Survey and Landuse Planning, Amravati Road, Nagpur, India

Kamalvir Tyagi

Reader, Department of Economics, M.M.H. College (C.C.S. University), Ghaziabad, U.P., India.

Kamta Prasad

Chairman, Institute for Resource Management and Economic Development, Delhi, India.

M. Istiaq

Professor of Geography, Jamia Millia Islamia, New Delhi, India.

M.R. Goswamy

Department of Geography, Jamia Millia Islamia, New Delhi, India

M.S. Sekhon

Department of Geography, DAV College, Chandigarh, India.

M.S. Swaminathan

Chairman, M.S. Swaminathan Research Foundation, Taramani Institutional Area, Chennai, Tamil Nadu.

Manfred Domroes
Department of Geography, Research Unit for Recent Climate Change, Mainz University, Germany.

Mrityunjay Mohan Jha
Bhoovigyan Vikas Foundation (Earth Care Foundation), New Delhi.

Murari Lal
Research Officer, State Land Use Board, Planning Department, Government of Uttar Pradesh, Lucknow, India.

N. Charthad
Lecturar, V.N. Government Institute of Arts and Social Sciences, Nagpur, India

N.C. Gautam
Director, Centre for Land Use Management, 25-24/1, East Anandbagh, Malkagiri, Hyderabad, India.

N.K. Baghmar
Department of Geography, Ravishankar Shukla University, Raipur, Chhatisgarh, India

Nalin Kumar Singh
Research Associate, M.M.H. College, Ghaziabad, Uttar Pradesh, India

Noor Mohammad
Professor and Head, Department of Geography, Delhi School of Economics, University of Delhi, Delhi, India.

P. Vijaya Kumar
Central Research Institute for Dryland Agriculture, Hyderabad, India

P.C. Agarwal
Retd. Scientist (Eng. SE) NIC, Government of India, SA-89, Shastri Nagar, Ghaziabad-201 002, U.P., India.

Paramita Majumdar
Research Officer, Institute of Social Sciences, Vasant Kunj, New Delhi, India.

R.B. Singh
Department of Human Geography, Delhi School of Economics, University of Delhi, Delhi, India.

R.K. Sharma
Agricultural Economics Research Centre, University of Delhi, Delhi.

R.M. Johri

Department of Botany, Post Graduate Studies and Research, M.M.H. College, Ghaziabad, U.P. India

R.S. Goel

Director, Member Secretary, National Environmental Monitoring Committee for River Valley Projects, 412(S) Sewa Bhavan, R.K. Puram, New Delhi, India

Rocket Ibrahim

Associate Professor of Geography, Jamia Millia Ismaia, New Delhi, India.

S. Girgaonkar

Lecturer, L.A.D. College for Women, Shankar Nagar, Nagpur, India

S. Hasan Ansari

Reader, Department of Geography, M.D. University, Rohtak, Haryana, India.

S. Kumar

Research Scholar, Department of Botany, M.M.H. College, Ghaziabad, U.P., India.

S. Lata

Department of Botany, Post Graduate Studies and Research, M.M.H. College, Ghaziabad, U.P. India

S. Mittal

Ecophysiology Lab. Department of Botany, C.C.S. University, Meerut

S. Poonia

Ecophysiology Lab. Department of Botany, C.C.S. University, Meerut

S. Sharma

Ecophysiology Lab. Department of Botany, C.C.S. University, Meerut

S.N. Das

Chief Soil Survey Officer, AIS & LUS, Government of India, IARI Buildings, Pusa, New Delhi, India.

Sanjay Kumar Singh

Lecturer, P.G. Department of Geography, Tilak College, Auraiya, U.P., India.

Satyaveer Singh

Assistant Director, National Institute of Agricultural Marketing, Jaipur, Rajasthan, India.

Smt. Mridula Singh

Additional Director, State Land Use Board, U.P., Yojana Bhawan, Lucknow, India.

Suruchi Tyagi

Reader, Department of Botany, M.M.H. College, Ghaziabad, U.P., India.

Surya Kant

Department of Geogrpahy, Punjab University, Chandigarh, Punjab, India

Syeeda Khatoon

Lecturer, Department of Economics, M.M.H. College, Ghaziabad, U.P., India.

T.S. Chauhan

Lecturer, Department of Geography, University of Rajasthan, Jaipur, Rajasthan, India.

Tapeshwar Singh

Reader, Department of Geography, M.M.H. College, Ghaziabad, U.P., India.

Tony Binns

Reader, Department of Geography, School of African and Asian Studies, University of Sussex, Falmer Brighton, U.K.

Victor Roger Savage

Head, Department of Geography, Coordinator, Southeast Asia Studies Programme, National University of Singapore, Singapore.

Vijay K. Gupta

Department of Civil and Environmental Engineering, Cooperative Institute for Research in Environmental Sciences, University of Colorado, Boulder, CO 80309, U.S.A.

Vipul Singh

Leturer, Moti Lal Nehru College (University of Delhi), Benito Juarez Marg, New Delhi.

Vishwambhar Prasad Sati

Asst. Professor, Department of Geography, Government P.G. College, Shivpuri, M.P., India.

Y.S. Ramakrishna

Project Coordinator (Agri. Met.), Central Research Institute for Dryland Agriculture, Hyderabad, India.